Mining in the Lake Counties

16s.
(80p.)

Other "Dalesman" books on the Lake Counties:

ARNSIDE AND SILVERDALE
AROUND ULLSWATER
CENTRAL LAKELAND
CUMBRIAN COAST RAILWAYS
FOLK LORE OF THE LAKE COUNTIES
GHOSTS OF THE LAKE COUNTIES
GRANGE AND CARTMEL
GRASMERE AND THE WORDSWORTHS
THE JOHN PEEL STORY
KESWICK AND NORTHERN LAKELAND
LAKELAND COOKERY
LAKELAND GEOLOGY
LAKELAND THROUGH THE YEAR
LAKELAND WALLS
LEGENDS OF THE LAKE COUNTIES
MAIN LINE OVER SHAP
MOTORING IN THE LAKE COUNTIES
SETTLE – CARLISLE RAILWAY
THEY CAME TO THE LAKES
UNDERGROUND IN FURNESS
UPPER EDEN VALLEY

Printed and bound in Great Britain by
FRETWELL & BRIAN LTD.
Silsden, Nr. Keighley, Yorkshire.
PATENT APPLIED FOR

	From Area codes 01923 or 0208:	From the rest of Herts:
Renewals:	01923 471373	01438 737373
Enquiries:	01923 471333	01438 737333
Minicom:	01923 471599	01438 737599

Mining in the
Lake Counties

by

W. T. Shaw

DALESMAN PUBLISHING COMPANY LTD.,
CLAPHAM (YORKSHIRE),
VIA LANCASTER.
1970

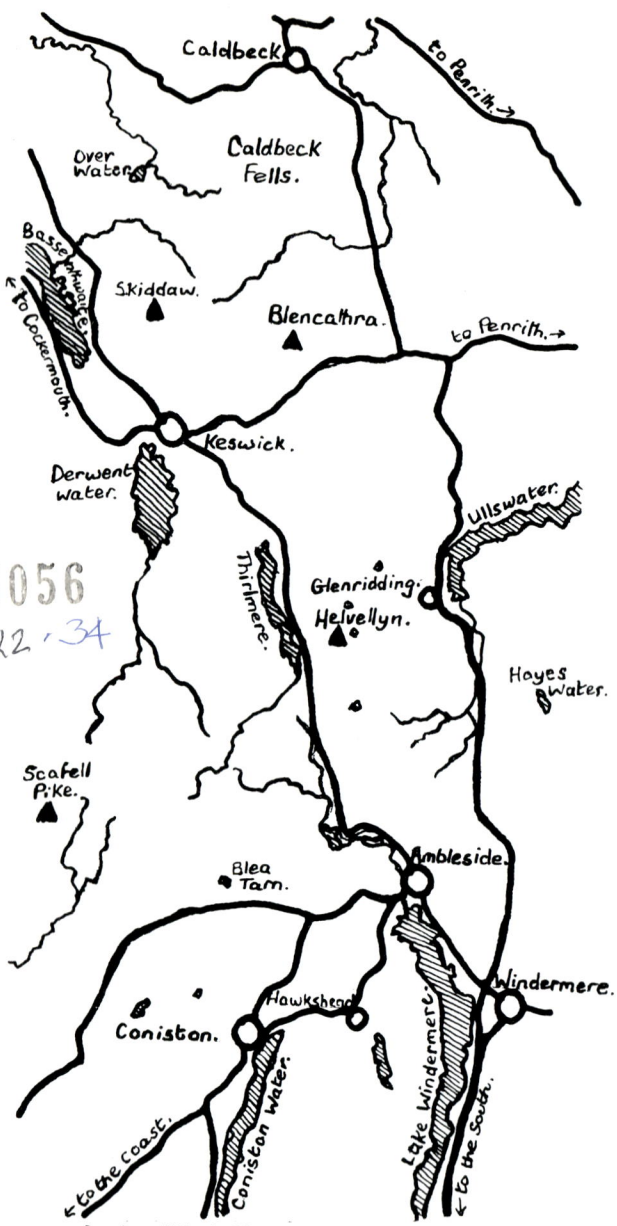

A simplified Map of the Lake District

4

Contents

The photographs appearing on pages 83 and 84 are from the collection of Mr. H. Oglethorpe, of Glenridding; photographs on pages 107 and 113 were taken by Mr. Joseph Hardman.

Cover photograph by Eric Holland shows the last shift at Sandbeds Mine, Caldbeck.

Introduction

*T*HE growing number of students and others interested in the geology, minerals and mining ventures of the Lake Counties prompts me to set down on paper some of the little-known facts of the once properous and important metaliferous mining industry.

Having worked in the mines of the District practically all my life, and being of an old Coniston mining family, I have made many friends in the industry and accumulated many notes and observations from associates now long deceased. As a boy I spent happy days at the trials on Tilberthwaite Fells with Joseph and John Hellen, two of the last copper miners, and I also prospected with "Uncle" John William Shaw, who was mine foreman at Coniston for many years.

I am grateful to many people who have helped me during this time. I remember particularly Captain Borlase and his sons, William and Edward, Mr. Anthony Wilson and Sir Arthur Russell Bart., all long deceased. Miss Jill Harvey and Dr. F. H. Day were with me on many geological excursions on the Caldbeck Fells. Mr. E. H. Shackleton has given useful advice, and McKechnie Brothers, Ltd. gave me permission to publish details of the Caldbeck Barytes Mines. I think also of the very generous help given by the archivists of the North Riding. The late Mr. Fawdry of the Leconfield Estates allowed me to browse through some of the old mining records. Many others have helped me in the preparation of this book.

W. T. SHAW

The Mining Story

Pre Elizabethan.

COPPER and iron mining in the Lake District go back at least to Roman times. Whether the inhabitants had any previous knowledge of mining and the production of metals would be difficult to prove, although it is possible that a number of the iron bloomeries relate to earlier times.

Some very ancient workings at the Coniston Copper mines are thought to date back at least to the Romans, and there is still to be seen a piece of donkey track beautifully chiselled out of a solid glaciated rock outcrop, near the entrance to the Bonsor Deep Level, which local legend attributes to them. As the Romans made roads and built camps throughout the district they would find signs of mineralisation at various places. Nothing remains elsewhere which is thought to be theirs. Lakeland then was thickly wooded and the valleys were mainly undrained. Travel was easiest by keeping as much as possible to the hills. The woods were largely composed of oak and birch, especially on the higher ground, and a small patch of this original forest can still be seen at Keskadale near Keswick.

Iron was in demand for making weapons and tools; the ore was carried to wherever there was a plentiful supply of charcoal for fuel and water power to blow the bellows. In some cases the ore was near at hand, as in Langdale and Grasmere, and the haematite deposits of Furness supplied the Coniston bloomeries, some of which are undoubtedly very ancient. At the foot of Glen Mary, near Coniston is the remains of a very ancient bloomery and there are small veins of haematite nearby but no sign of any workings have been found.

For centuries after the Romans left no records survive; and it is not until the Normans came that definite place names appeared, showing that some of the mines had been discovered long before. King Edward III made a grant of mines and appointed Robert de Barton to be keeper of silver lead mines at Silver Beck and Minersdale and copper mines at Keswick. There is little doubt that Silver Beck is Silver Gill Mine and Minersdale is thought to be the ancient name of the valley in which stands the famous old Roughtongill Mine on the Caldbeck Fells. Of the copper mines near Keswick the Goldscope is the most ancient having been considerably worked before Queen Elizabeth I.

King Edward IV in 1474 appointed a commission to inquire into the mines at Alston and Keswick, which were apparently then being worked,

MINING IN THE LAKE COUNTIES

and when the German mining expert Daniel Hechstetter came to Keswick for Queen Elizabeth I in 1565 there were plenty of old workings for him to examine.

Elizabethan.

Hechstetter was a learned and able mining engineer; he set about establishing a copper mining and smelting enterprise on a scale without equal in England in a district far from any industrial facilities and without any roads as we know them. Local miners were few, and as a large number were required he succeeded in persuading many Germans to come over from the long-established mining districts of Tyrol.

At first there was friction with the locals who thought the Germans, or Dutchmen as they were called, would keep the work for themselves. Soon there seems to have been plenty of work for all who wished and some Dutchmen married local girls. Such an invasion by outsiders was repeated at Coniston about the 1840s when many Irish families came to work at the mines but intermarriage quickly put an end to the disagreements.

Hechstetter undertook an enormous task even when surrounded by his own people. Few stores existed at Keswick and they had to come by sea from afar to Newcastle or Workington; they were hauled to the various sites over almost non-existent roads. It was easier to carry things by packhorses than on wheeled vehicles and only the heaviest parts of the machinery were carted.

The site for the Keswick smelter, Brigham beside the river Greta, had a plentiful supply of water. Here was constructed a cluster of furnaces with refinery and eventually a rolling mill, all of the best design then available. The Goldscope mine was reopened and developed along with the Dalehead and Copper Plate mines. At the same time the Caldbeck mines were reopened and equipped with a crushing and dressing plant. Grasmere mine was started and here a dressing plant was built. Much ore was raised at Coniston and carried by packhorses to the Keswick smelter.

The quantity of peat, charcoal and stone coal devoured was for those days enormous and dozens of men were employed in carrying these materials; others were engaged in felling wood and making it into charcoal, cutting and drying peats and mining the stone coal. This latter came from Caldbeck and Bolton, near Wigton, where the Company opened their own collieries. Some of the charcoal was made locally but a great deal was brought from Furness, where organised wood cutting was already established for this purpose.

The coppice woods, as they were called, were cut on a rota system of about 18 to 20 years. The best standard in a clump of small trees was very often left to grow and in this way many of the large oak trees came into being. Much of the peat was cut behind Skiddaw and also near Mungrisdale, where there is still a good deposit. The local people obtained peat for their fires within the last 30 years. So much oak wood was used for construction works and in the mines that there was a very real shortage of good trees in the district and the Company thought of bringing some from Ireland.

An impression of Elizabethan miners washing ore.

At first the lead used in the copper refinery for cupelling the silver was brought from Alston Moor, which shows that none of the local lead mines were in work. No doubt Grasmere was started for this purpose; at first it must have looked very promising but it was very disappointing. Stoneycroft mine was either started or restarted without much success and some years elapsed before Brandlehow, Yewthwaite, Barrow or Thornthwaite were tried. Meanwhile Roughtongill, on Caldbeck Fells, kept up a steady supply.

The Greenside mine had apparently not then been discovered, which is not surprising as it does not seem there would have been much of an outcrop showing on the green mountainside and it was most likely found only by remote chance.

17th and 18th Centuries.

The famous black lead (plumbago) mine at Borrowdale, which had long been discovered and worked on a small scale, passed into the hands of the Banks family early in the 17th century and the pencil-making industry at Keswick was established.

Oliver Cromwell's men are supposed to have been responsible for the destruction of the copper smelter and mining works in 1651, a senseless act from which the copper enterprise never really recovered.

Even by then most of the best ore had been won from Goldscope and the smelter relied more and more on ore from Coniston and Caldbeck for the other Keswick mines were not nearly so rich.

Nearly 40 years were to pass away before there was a revival. In 1690 another party of "Dutch" adventurers came to the district and rebuilt the copper works. Whether these people were Germans or really Dutch is not known but they also built a lead smelter in Stoneycroft Gill and either started or reopened the Greenside mine. Coniston mine was worked by the Le Flemmings with some vigour and the mining industry as a whole began to take on some of its former glory.

By now iron production had become of more importance than formerly and large furnaces sprang up at Langdale, in the Coniston district and at Langstrath near Borrowdale. Much iron ore was won at Red Tarn (near Pike o Blisco), Ewer Gap and at Tongue Gill and it is said that the iron made at Langdale "was not much inferior to Dantzig."

During the 18th century there was a good deal of mining with lead rather than copper, the predominant metal sought after. Coniston remained the most successful copper mine and was probably worked more or less continuously. The Keswick copper mines were but little worked and the smelter came to an end fairly early in the century. The lead mines were worked with more vigour and kept the Stoneycroft smelter in continuous work. On the Caldbeck Fells, a smelter was at work at Roughtongill mine, and at the Hartsop mine in Patterdale there was also a small smelter.

The lead ore from Greenside was carried on horseback over Sticks Pass to Stoneycroft but it is not thought that the mine was worked continuously.

19th Century.

The 19th century brought the greatest mining activity in the district since the coming of the Elizabethans. There was an improved road system and the passage of carts and horse-drawn lorries became easier than formerly. Soon the packhorse system of transport came to an end.

Over the centuries much use had been made of the Lakes for transporting ore and materials by boat, and this was especially so on Coniston and Ullswater. To a lesser extent ore and materials had been ferried across Derwentwater to Keswick rather than carted or carried around the shore. At Patterdale there were flat-bottomed boats which could carry up to eight tons and these were sailed for nine miles on Ullswater with loads of lead ore but more regularly with roofing slates from the then prosperous Hartsop slate quarries. At Coniston the same system was in vogue with extensive lakeside quays constructed at Waterhead and Nibthwaite. The bulk of the copper ore was transported in this way, being carted from the mines to Kirby Quay, loaded into boats and sailed down the six mile lake to Nibthwaite. Here it was loaded in to carts and taken to Greenodd for shipment to South Wales smelters.

This system of transport came to an end at Coniston when the Coniston Mining Company constructed the branch line from Foxfield to Coniston, a distance of about 10 miles. It was completed in 1865 and was afterwards taken over by the Furness Railway Company. The mining company's original locomotive, which was christened "Copper Nob" is preserved in the railway museum. Soon after the railway came to Penrith the boat haulage on Ullswater also came to an end. The very considerable traffic of lead from, and coals to, Greenside mine ensured that the road to Penrith was always kept in a reasonable state. By 1867 the railway had reached Keswick and there again this made a great saving in the distances to be travelled by horse and cart.

As the county's railway system was developed, the Lakeland iron furnaces came to an end because iron ore was available in West Cumberland and Furness, with nearby supplies of stone coal and steam power, so that it was no longer necessary to build the furnaces where there was water. The use of charcoal for fuel became almost extinct. Nevertheless at Backbarrow, near Newby Bridge, the old-time iron-works continued to smelt ore from Lindal Moor using charcoal produced from the coppice woods in the area between Coniston and Windermere lakes.

Early in the century the London firm of John Taylor and Son took over the management of the Coniston mines and opened them up on a grand scale, so that for over 50 years they employed great numbers of men and boys, with a maximum said to be 600. At Patterdale the Greenside Mining Company was formed and the mine developed into a first class operation which employed as many as 300 men and boys. The Threlkeld mine was taken up by Captain William Bawden in 1879 and by 1880 employed 200 men and boys. All over the district there was great activity and much prospecting went on. At Keswick the great lead bunch was found at Goldscope mine and this company also worked the nearby Yewthwaite lead mine.

Brandlehow mine was in full production and employed not less than 100 men. In a smaller way work was going on at Barrow mine where a new shaft was sunk towards the end of the century. The Force Crag mine was worked for 30 years for lead. Thornthwaite mine was modernised and set going by Captain William Francis on a greater scale than ever before, and here at least 100 men and boys were employed.

The Patterdale Mining Company made a valiant though unsuccessful attempt to re-establish lead mining in the valley outside Greenside. The Helvellyn mine was discovered and a Company formed to open it out.

On Caldbeck Fells the Roughtongill and Driggith mines were working at full capacity though both became worked out in the 1870s. In 1874 there were 12 lead mines, three copper mines, the plumbago mine and three iron mines in production in the Lake District. At many more prospecting and development work were taking place.

Towards the end of the century, the large discoveries of metals in Australia, Canada, U.S.A. and Spain brought the prices tumbling to well below the economic limit of most local mines; some managed to survive but they were greatly curtailed, and became shadows of their former glory. Only Coniston among the copper mines lingered on, growing steadily less year by year as development work ceased and the reserve ore was extracted.

Greenside mine was reorganised and modernised with the introduction of electrical power and the economies in working costs kept it going on the same output scale with a greatly reduced labour force. On Caldbeck Fells metal-mining ceased in 1878 when Roughtongill closed for good, to all intents and purposes exhausted despite a great deal of expenditure on development all of which proved abortive. Threlkeld mine continued on a rather smaller scale but Thornthwaite was worked with great vigour by Mr. Anthony Wilson, who succeeded Captain Francis as manager. Brandlehow and Barrow mines came to an end about 1890 for with the death of Mr. Vercoe who was their able mining engineer and the low price of lead, the company became dispirited.

The heyday of Lakeland mining was past; a new factor was beginning to appear and it had no relation to the richness or poorness of the veins or prices of metals. This was the effect of mining works on the scenery. To some extent the falling off in importance of the mining industry to the community helped to foster this opposition for instead of some 2,000 men and boys directly employed at the mines in previous years there were now less than 500.

20th Century.

The 20th century has brought two world wars, neither of which played any great part in reviving the industry. Coniston Mines had a brief period of prosperity at the beginning of the 1914-18 war when a new electrolytic copper plant was constructed, but the loss of the principal engineer, Count Henri de Varney, who was killed when his plane was shot down by the Germans in 1915, brought the works to an end. Apart from a little prospecting work in 1930 and 1954 nothing further has been done.

A process for the extraction of copper from the extensive waste heaps is now being tested and if successful these may be treated in an endeavour to extract the little copper thrown away by the old miners. The chances of the mines being reopened are considered extremely slim, the extensive old workings being now in a ruinous condition. The amount of money required to open out new workings to get below the "old men" is very considerable and unfortunately must be laid out before any real estimate of the quantity of copper likely to be obtained can be made.

The great Greenside mine was completely worked out by 1962 and with its closure came the end of lead-mining at Patterdale, for although other likely veins exist, the opposition to their being worked makes it not worthwhile to try. On Caldbeck Fells the Barytes Mines were continued on veins including some which had long been known to the ancients. Because they contained no lead or copper of value they were left unworked, for barytes was of little or no value in the 19th century. In 1940 McKechnie Brothers Limited, of Widnes, took over these mines and for 26 years they were worked with success until all the worthwhile ore was exhausted and once again mining on these fells came to an end.

In the 1914-18 War the Carrock Wolfram Mine was greatly worked with over 100 men employed for most of the duration. Wolfram is a mineral which has great uses in the making of armaments and so in peacetime is generally in only slight demand with resulting greatly reduced price.

The slump in metal prices at the end of the 1914-18 war brought to an end Force Crag, Thornthwaite and eventually Threlkeld. Since then the barytes deposit at Force Crag has been worked considerably and at present the mine is being prospected in the hope of producing lead and zinc. It has the doubtful distinction of being Lakeland's only active mine.

The total number of people employed in metal mining in the Lake District is now six, which shows the astonishing decline in the industry in the last few years. The reasons for this are not far to seek. 1. The difficulty in obtaining planning permission has stifled attempts at a revival; 2. There has been opposition from the various scenery protection societies; 3. Large sums of speculative capital are required nowadays in mining ventures; and 4. The taxation system does not allow initial capital to be recouped before any tax should be payable.

No. 0 Level at Force Crag Mine, 1969.

13

Lives of the Miners

THE old-time miner dressed in thick wollen trousers, heavy flannel shirt and a long leather jacket. He wore clogs and had a battered old felt hat, items which survived until rubber boots and safety helmets became general. The woollen trousers gave way to fustian and the leather jacket was eventually replaced by any old ordinary jacket with a waterproof coat for use in wet places. The leather coat had short sleeves and was fastened across the chest by leather thongs. The long back came to a rounded point so that when the wearer bent at his work the water dropping on him ran off clear of his legs.

Illumination was by candles, oil or tallow lamps and it appears that in Elizabethen times tallow lamps were in common use. These were replaced by tallow candles, or "dips" as they were called, in either the 17th or 18th century, and modern acetylene lamps did not come into general use until after the 1914-18 war. Electric lamps have not been used in the Caldbeck Fell mines but were used at Greenside Mine during its last years. The fumes and smoke off the tallow lamps can be well imagined especially in a badly ventilated working place and no doubt most of the local tallow would be from sheep.

The lamp itself was a simple metal vessel with a spout through which the thick wick was threaded. A hook was provided for carrying purposes and also for hanging the lamp on the rock or timbers. The flame heated the vessel sufficiently to keep the wick supplied with liquid tallow. The tallow candles or dips were supplied to the mines in large bunches joined together by the wicks; they could be hung over a metal ring which was drawn up by a rope to near the ceiling of the stores to prevent the rats from feeding on the candles. Each man was allotted so many candles per week depending on the state of the ventilation where he worked. At some mines the quantity was a standard amount and the miner was entitled to take home any candles not used. The candles were generally set in a ball of clay which was then stuck on any convenient place in the working and was also often put on the front of the old felt hat. Miners' candlesticks in general use in the 19th century were probably almost as old as the use of candles.

A form of glassless lantern known as a "bug" was popular for use in draughty places and also when wheeling barrows or tramming. It was made from a piece of sheet iron about 9″ high fitted to a wooden top and bottom about 5″ diameter and 1″ thick. There was an opening at the front of roughly 4″ wide x 9″ high. A hole was bored in the centre of the bottom in such a way that a candle was a good tight fit. A candle was pushed up through about 2″ above the bottom and was then further advanced as it burned away. A strap was fixed to the top of the lantern for carrying purposes. When wheeling barrows or tramming bogies it was usual to make a hole in the top of the load for the lantern to sit steady.

The old men wore footless stockings with their clogs, which were lined with soft dry straw cut to the right length. If the wearer trod in water over the top of his clogs he just took them off, emptied them out and replaced the wet straw with some dry which was always carried for the purpose. Thus to get wet feet was not such a problem as in later days with footed socks and stockings. I remember my grandfather cutting clog straw; he wore ordinary stockings and so the straw must have been for extra warmth.

Every village had a "clogger" in those days and there is still one at Caldbeck. Usually he made his own soles from birch or "eller", more rarely of beech, and he always had a stock of all sizes partially finished so that with a few more deft strokes with the double hand knife a perfect fit was assured for each wearer. To make the soles last longer amongst the sharp stones the miners had them heavily caukered and the main irons were a type which reached right up to the heel. Inside were fitted rather smaller irons and, inside these were fitted strips of caukering or heel caukers. If the wearer's name began with S they were often set so that he left his initial printed in the clay as he walked. Children's size caukers were fitted inside the heel caukers. In sulphurous mines clogs were put together with brass or copper nails and in the Threlkeld mine the water was so acidic that a pair of ordinary clogs only lasted six weeks.

Boots were rarely worn; they were more costly, did not last very long among the sharp rocks and had to be dried more carefully than clogs. One drawback to the iron-filled clog soles was that in soft snow large "patterns" were picked up and the wearer had to constantly knock them off against the wall or wayside boulders. Nowadays very few men wear clogs and metal-miners wear rubber Wellingtons, some of which have special toecaps inserted as a safety measure.

The long leather coat disappeared with the thick flannel shirt and fustian trousers and the miners of today wear ordinary shirts, a pullover of some sort, very often terylene trousers, Wellingtons and a safety helmet with an ordinary jacket augmented by an oilskin coat and leggings for wet places.

Few beards are now seen among miners, but in the not-so-olden days they were very common. I can remember many of the old miners who had beards or moustachies. Some thought that the wearing of whiskers was a protection against the ravages of dust but there is no evidence that this was so for at Patterdale many died from silicosis whether they had beards or not. All the lead and copper-mining on Caldbeck Fells had come to an end before the introduction of dynamite and at Coniston by far the greater part of the mining was done by gunpowder and hand drills. These miners were less affected by this dread disease than their contemporaries at the Greenside Mine.

MINING IN THE LAKE COUNTIES

During most of this century the use of water through the boring machines, and of air-driven fans to remove the smoke and dust from the blasting, have greatly improved working conditions compared with those prevalent before the 1914-18 war.

Hours of work are much the same as in Elizabethan times with the usual mine shift of eight hours. Even in those days five days a week were worked in the mines, for it had long been recognised that miners did not generally live as long as farmers and other outside workers and a short day and week were considered long enough to toil in the bowels of the earth.

The wages paid to miners have always been rather higher than those of agricultural workers and by taking contract, or "tut work" as it was called, it was often possible to make a good deal more than the ordinary daily wage. The following gives some idea of the day rate wages of the miners over the ages in Lakeland.

1569	*English miners at Roughtongill mine*	*6 pence per day*
1858	*Miners at Roughtongill mine averaged*	*3/- per day*
1858	*Miners at Coniston mines*	*2/9 per day*
1869	*Miners at Coniston mines*	*3/10 per day*
1870	*Miners at Driggith mine on special work*	*4/- per day*
1900	*Miners at Greenside mine on day work*	*4/- per day*
1928	*Miners at Caldbeck barytes mines*	*8/- per day*
1929	*Miners at Greenside mine*	*10/- per day*
1940	*Miners at Caldbeck barytes mines*	*9/9 per day*
1966	*Miners at Caldbeck barytes mines*	*48/9 per day*
1969	*Miners at Force Crag mine*	*60/- per day*
1969	*Miners at Force Crag mine on contract*	*100/- per day*

Contracts were usually let by the month or in some mines in three-monthly periods. The mine manager or captain kept a book in which the measurements of advance in each working place were recorded, as they were measured at the end of each period When a new party was to work in driving levels a "mark" was made by boring a hole about three inches deep, about shoulder high, on the side of the level a little way back from the face. This distance was then carefully measured and noted in the book. When measuring day came around the place was measured from the borehole and the mark was deducted to give the actual advance of the miners.

Progress was so slow in the handwork days there was little likelihood of the manager getting mixed up with the amount of work done in the period in individual working places, but nowadays it is usual to measure up weekly or it would be difficult to keep count of all the work done. Whereas the old miners cut only about one foot of level per man per week in hard ground, latterly two men have regularly cut 25 feet per week in similar rock, and in the soft oxidised barytes ground well over 100 feet per week has been achieved.

The contractor paid for the miner's explosives, tools and sharpening from his price per fathom. If a miner failed to make a wage he was left to the generosity of the manager, for there was no make-up wage in those days, nor indeed until the 1930s at Patterdale. In the barytes mines the miners on contract were paid by the foot with set prices for each type of ground encountered and there was no question of the price being cut if the men managed to make good wages; the more ground that they cut the better it was for the company.

Miners in Hartsop Mine, 1932.

17

MINING IN THE LAKE COUNTIES

Most of the mining work has been performed on contract and as there are few records of latterday tributing it seems unlikely that this system of working was much practised. The Dutchmen used it in all their workings and, in their day mixed contracts of both measurement work and tonnage were quite common. In tribute working, the tributer is in fact a miner who is paid not for the amount of ground he excavates but for the finished ore he raises from it. At Coniston mines up to the end of the 19th century the company, or adventurers as they were often called, worked all the most profitable ore ground by direct measurement contracts.

Side veins and the poorer ore ground were the home of the tributers who contracted with the company to work them, being paid for the actual dressed ore produced. Depending on the prospects and richness of the tribute pitch, the price would be reckoned at so many shillings in the pound of the value of the ore raised. Very often in the Coniston mines this was 15/– but there are instances of only 5/– in the pound being paid. The tributer was virtually his own master and followed the vein as he thought fit, coming and going to his work as he pleased. All powder, candles and tools had to be paid for and there were usually charges per ton of ore for drawing and dressing it; each tributer's material was treated separately.

It was common practice for the very poor portion, or stamp ore, to be claimed by the company on the contention that it scarcely paid to treat it and that its worth to the tributer after paying the dressing cost would often as not be a debt. Some families at Coniston worked only on tribute, scorning "tut" or day work, and though they sometimes did well it is improbable that their wages over the years equalled those of a good "tut" man. There is little evidence of the tribute system on Caldbeck Fells after the days of the Dutchmen but prior to the 1914-18 war there was some at Carrock Wolfram mine.

It takes years to become a skilled miner, a fact all too often overlooked by people in authority who are apt to class miners as labourers instead of the skilled men they are, and sadly one watches the metal mines close one by one with the skilled know-how being lost for good as the miners drift into other occupations. My father said that by the time a man got to be a really good miner he was physically past it, and this saying illustrated the long period required to acquire the know-how to deal with all the mining problems. As no two mines are alike a man had to be versatile to be equally at home in any ground. In the not so long ago when mining in Lakeland was still an important industry, the miners often moved from mine to mine in an endeavour to earn higher wages; thus they became much more generally skilled than they would have been had they worked always at one mine.

Methods Used in the Mines

Breaking Ground.

MINING practice by the ancients changed little from one century to another but a great deal of skill was required to make any sort of progress using hammers and wedges in hard solid rock. It is said that in those days a miner served a seven years apprenticeship.

When the Germans or Dutchmen as they were called came to Lakeland for Queen Elizabeth I they brought with them new ideas, new and better tools and were able to cut their way through the solid rock more easily than had been previously possible. As the wages of the Dutchmen were very superior to those paid the English workers the latter would no doubt be anxious to learn what they could from the foreigners and so be able to earn more money.

The Dutchmen used a variety of wedges in their work, some small some with a wooden shaft and some quite large for use with the heaviest hammers. The wedges with the shaft attached were held against the rock with one hand and struck with a single hand hammer in the other. On opening up a small crack a stronger wedge was hammered in and driven home with the heavy hammer. In reality, in really solid rock, it must have meant virtually cutting it away as with chisels and would, indeed be a laborious process. Many years ago men broke through into an ancient pre- gunpowder shaft in the Hartsop lead mine and found a small wedge the handle eye of which was broken. This tool would be at least 300 years old.

The ancients also used a useful tool which seems to have been discarded a long time ago. This was a combined pick and "corrack", or rake, and was probably greatly used when making "hushes". Single-pointed or "poll" picks were used by the old men, for the hammer end served to drive wedges as well as to break off pieces of rock. Detachable pick-heads did not come into use until near the end of the 19th century and all previous ones had eyes similar to a hammer head and were fixed permanently to the shaft. When they were being sharpened the blacksmith had to be careful not to dry out the shaft and so make the head slack.

All the old tools were made of the very best iron, generally from Low Moor, and were tipped or faced only with steel. At any mine there was a great deal of blacksmith work and a good smith was a valuable man. The correct tempering of the tools was extremely important.

MINING IN THE LAKE COUNTIES

In the oxidised upper rocks on Caldbeck Fells the pre-gunpowder men would be able to make reasonably good progress through mostly soft rocks. The large patches of hard solid quartz which occur in the veins would be byepassed unless they contained any valuable ore. Unfortunately none of the old workings are open at the present time but there is a record of some of them being entered from the upper part of Roughtongill mine, and the Old Dutch level at Red Gill mine is still probably standing open away from an entrance which long ago collapsed.

The use of firesetting was not general in Lakeland and there is no evidence of its use on Caldbeck Fells. It is known that the Dutchmen worked open stopes or "nicks" on Coniston Fells by this method but nothing about its use underground has been recorded. Briefly brushwood and small faggots were heaped along the floor of the working, which had been previously cleaned or "swept" of all loose material, and then it was set on fire. When only the red embers remained, a man who was suitably protected by thick leather apparel threw on a mixture of vinegar and water and with a great hissing of steam the rock was cracked and rent, small pieces flying like shrapnel. The loosened rock was then all wedged out and the nick swept clean ready for the process to be repeated. It is often said that there is nothing really new, and present day jet-piercing is a modern form of fire-setting.

Quicklime had been used with some success for rending rocks long before the use of gunpowder. Its use was limited and only down holes could be charged in this way. Being slow in action, only fairly soft rocks could be tackled. Near Tarn Hows at Coniston about 50 years ago the remains of holes blasted in this way could be seen with the lime still adhering to them. Apparently after a hole was bored in the rock it was about half filled with quicklime and securely tamped with stemming material around a pricker or needle (similar to that used later in gunpowder blasting), it was then withdrawn and the long thin spout of a special funnel inserted. Some water was poured down and after the funnel was withdrawn as quickly as possible a final tamp was given. The operator retired a safe distance to watch whether the rock gave way or the stemming merely blew out.

The date of the first use of gunpowder for blasting in mines is open to argument but legend has it that the Dutchmen were using it at Coniston early in the 17th century. One of them, Simon Puthparker, was killed in a blasting accident in the stope or nick near Leverswater which still bears his name (Simon's Nick). If they were using gunpowder at Coniston then there is no doubt it would also be in use in their other mines, a tremendous step forward in the art of breaking ground. There was to be little further progress for the next 300 years. With the use of gunpowder the art of wedging and chiselling a tunnel forward soon disappeared and was lost forever.

Boring the shot holes now became of paramount importance and both single and double hand boring was practised. In single-hand boring the miner held the drill steel, or "jumper", in one hand and struck it with a 3 or 4 lb hammer, lifting and slightly turning the steel after each blow. Men were either right hand or left hand and few were truly

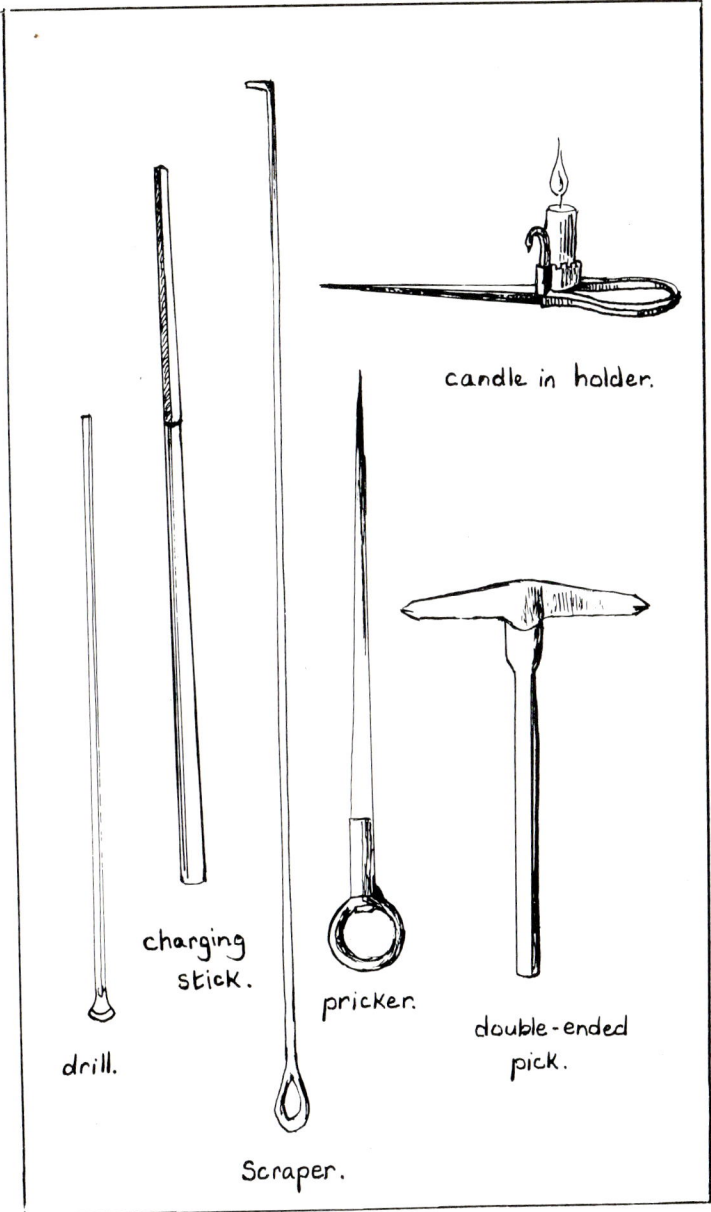

candle in holder.

charging
stick.

pricker.

double-ended
pick.

drill.

Scraper.

A collection of mining tools.

ambidextrous. Left hand men were often scarce and as they were needed to keep the holes parallel on the left side of the working they were some times paid slightly more if they were "naturals".

Boring in some mines was performed by two men, one turning the jumper and the other striking with a much heavier doublehand hammer and they usually worked spell about. In very hard ground it was proved that two single-hand men could bore more footage per shift than double-hand men and they did not blunt so many jumpers but in medium hard rock the double-hand men usually won especially when boring breast and overshoulder holes.

As many as possible of the holes were drilled downwards so that water could be added, not primarily to keep down the dust but because most rock bores faster with water. A pad or "blasher", usually of old sacking was fitted over the jumper to prevent undue splashing and water was poured in from a can as required.

Boring hammers were from 3 to 4 lbs. weight for single-hand work and up to 7 or 8 lbs for double-hand. They had small faces, the idea being that a hammer with a small face gave a sharper and therefore better cutting blow than the dull impact from the broad faced type. Before Bessemer steel became common all the jumpers, picks, hammers and wedges were made of best Low Moor iron and were tipped or faced with steel. Eventually all steel tools took over and a great deal of skilled blacksmith work was not needed.

The jumpers were usually $\frac{3}{4}''$ diameter in either round or octagonal section and were in various lengths from about a foot upwards. In most hard-rock mines a 3 feet jumper would rarely be used. Nearly all the holes appear to have been under 2 feet in length. The jumpers were sharpened or "bitted" by heating and forging to a chisel-like shape, being slightly curved on the cutting edge, the angles of which varied according to the hardness of the rock to be bored. For very hard rock the angle was greater than that used to drill softer rock. Shoulders were made which gave clearance as the bit wore; the sizes of the bits decreased as the jumpers lengthened so that they would follow each other as the hole deepened.

In driving tunnels, or levels, the Elizabethans and their predecessors always dug out the bottom of the face first and broke the top portion down last. The gunpowder men did exactly the same and several of their working faces have been seen in Driggith Mine, where all the holes bored could be observed and counted. By following this method the maximum number of holes in which water could be used was drilled and the minimum number of over-shoulder and "square" holes were required. Both over-shoulder and square-boring are the hardest work because the weight of the hammer has to be carried all the time and gives no help in striking the blow.

The old men hated putting up rises even though there was then no water trouble, and the broken material would fall down when blasted because boring up holes was very hard work both in striking and turning the jumper. They preferred to put down small shafts or "sumps" to connect one level with another even though there was generally a lot of water to wind as well as the broken material.

METHODS USED IN THE MINES

Before the advent of safety fuses in the 19th century the holes were fired by means of a straw or reed filled with a special fine gunpowder and ignited by a piece of touchpaper. These straws were usually of oats and the length depended on the growth of the season because they were of use only from the ground to the first joint, They were carefully cut, tied in bundles and hung up to dry in the miners' kitchens for each miner supplied his own straws and touchpaper, the latter prepared by soaking unglazed brown paper in a saturated solution of saltpetre. Each batch was carefully tested for running speed before being taken into the mine!

The gunpowder used on Caldbeck Fells more than likely came from Elterwater Powder Works, the nearest source, and carting it had problems. In the old days at Greenside Mine they had a special powder cart which had brass hubs and was still being used for general purposes some 40 years ago. The gunpowder was usually supplied in 25 lbs. capacity wooden casks which were made at Elterwater from local hardwoods and were bound together by hazel "withies", or bands; no nails were used. The gunpowder lay in a waxed bag and generally both bags and casks were returned when empty.

The C.B. grain was used for blasting but very fine powder similar to that used in sporting guns before the invention of smokeless varieties, was supplied for the straws.

It is not known if safety fuse was used in either Roughtongill or Driggith Mines; it may well have been because it was in general use at Coniston Mines in 1872, when a 24 foot coil cost 4d.

Prior to the Metalliferous Mines Act of 1872 it was usual for each party of miners to take a 25 lbs. cask of powder into the workings and place it in a safe dry place some distance from the blasting operations. Many wooden ends of these casks were turned out when old workings were cleared. After 1872 all gunpowder had to be carried underground in special canisters of not more than 4 lb. capacity, and as each party of miners had to purchase these regulation canisters from the mining company there would be opposition to the new regulations.

To charge and fire the drill hole was a great skill. It had to be cleaned and dried out using an iron scraper, through the eye of which was inserted a piece of old sacking. Then the gunpowder was put in place using a cartridge stick. This was a round stick about 3 feet long with a piece of copper or brass tubing, about 1 ″ diameter and a foot long, fitted on one end. The tube was filled with gunpowder from a copper bottle or canister and pushed into the hole bottom. A quick flick emptied out the powder and the process was repeated until there was deemed to be sufficient in the hole, usually approximately half its depth.

A small lump of clay-stemming was then inserted and pushed on to the powder. The pricker was then pushed home through this stemming and into the powder. More stemming was inserted and hammered tight using the stemmer. These tools were originally made of iron but were condemned by the 1872 Mines Act and after this date were made of brass or copper. These metals would not produce a spark. As the stemming was beaten into the hole the pricker was eased slightly to avoid

MINING IN THE LAKE COUNTIES

it being fastened in and also to let out any air which was being compressed. This was very important; many premature blasts were thought to have been caused by air compressed in this way and made hot enough to fire the powder.

The stemming material was often obtained from cross courses or faults and was a stiff claylike material. In mines where barytes occurred the mineral was crushed into a coarse sand and was greatly used as stemming for it would not strike fire and it became so dense when the explosion took place that it still remains in the old shot holes in Driggith Mine after over 130 years. When the stemming operation was completed the pricker, a tapered copper needle with an iron eye, was withdrawn and a straw inserted down the small hole. A wisp of touchpaper was twisted around the head of the straw and the shot was ready to be fired.

Seldom were more than two holes fired in a face at any one time and each hole was placed to be totally independent of its neighbour. Straws were still in use in the Coniston slate quarries when I was a boy and I have actually fired holes by this method. When one considers the millions of holes which have been blasted by straws, the small number of reported accidents was remarkable.

Three inventions in the 19th century brought about almost a revolution in mining methods. These were the use of safety fuse, compressed air driven rock drills and dynamite. The first use of rock drills in Lakeland was in the Dunmail Raise tunnel for Manchester Corporation's Thirlmere waterworks scheme. The directors of Greenside Mining Company paid a visit to this tunnel and were so impressed with the progress being made that they immediately installed a compressor and boring machines at Greenside.

The first rock drills were built on the principle of a steam engine cylinder and piston;they incorporated a piston some 4 inches in diameter which had a stroke of some 6 or 8 inches to which the drill steels were securely clamped. Thus they were hammered into the rock at each forward stroke and were withdrawn on the return stroke, during which they were partly rotated the whole action being similar to that of hand-boring. Built mainly of cast and wrought iron these machines generally weighed $1\frac{1}{2}$ hundredweights and were supported by a rigid screw bar set up either vertical or across the level and fed forward, as the hole deepened, by means of a handle turned feed screw. The machines worked on compressed air at a pressure rarely exceeding 60 lb. per sq. inch but it was possible to penetrate the hardest rock at a speed previously unknown. Rock-drills were installed at Coniston and Thornthwaite about 1890 and heralded the end of the hand-boring which had been the mainstay of operations for nearly 300 years.

Safety fuse was little used before 1874, when dynamite came into regular use. Because of the sensitive nature of the dynamite of those days there were several serious accidents, some of them fatal, but it was made safer to handle, especially in frosty weather when non-freezing varieties were invented. Even where hand-boring was still being done the use of dynamite cut down the number of holes required by at least half and so gunpowder was for mining a thing of the past just as it had superseded hammer and wedges 250 years previously.

24

METHODS USED IN THE MINES

The drill steels for the early rock drills were generally 1 inch diameter round in section and were hand-sharpened and similar in shape to a hand drill. Machines were invented which could forge cross and other type bits. About the time of the 1914-18 war a different type of rock-drill came on the market and was the forerunner of those in use today. In these machines the drill steel was not secured to the piston but fitted loosely into the front or chuck, where it was struck and rotated by the free moving piston. This was even more akin in action to a man hand-boring. The machines required a higher air pressure to enable the best results, for the greater the air pressure the greater the number and weight of blows per minute and thus the greater the cutting speed. By fitting a small water tube through these machines and using hollow drill steels a small stream of high pressure water could be fed down the holes being bored and at last it became possible to eliminate the harmful dust produced by machine-drilling, for the large numbers of miners affected by silicocis was soon all too apparent.

The lead mines on Caldbeck Fells ceased before either dynamite or rock-drills were used but in the barytes mines both were exclusively employed. Most of the work, especially in the rotten ground associated with the workable barytes deposits, was done using small hand-held machines.

At Greenside after the 1939-45 war a new type of device for holding and pushing heavy hand-held types of rock-drills was introduced from Sweden and there were drill steels tipped with tungsten carbide. This combination proved to be the greatest step forward in mining since gunpowder, for even in the hardest rock it was possible to bore holes at speeds of around 1 foot per minute. The same drill steel could be used for a number of holes instead of having as many as half a dozen ordinary drill steels to bore one hole.

This system has become standard practice throughout the mining world and at Force Crag Mine today a regular 5 feet per day is being advanced in the adit level by this method. In this ground 100 years ago the hard-worked miners would have done little more than 1 foot per man per week. Thus in the 400 years since the arrival of the Elizabethans to Lakeland the progress in the art of breaking ground has been sensational and the methods and results of today would probably have sounded more fantastic then than the prospects of mining on the moon are today.

Ventilation and Drainage.

TWO vital problems, ventilation and drainage have faced the miners and both can be costly to accomplish. It was long ago discovered that if water was led to the top of a vertical pipe or square box fixed in a shaft and let fall freely the entrained air could be separated from the water in a baffle box at the bottom and, as it was then under some pressure, it could be passed through square wooden boxes or metal tubing for quite long distances and so bring fresh air to an otherwise unworkable face. Several of these ventilators, which were called "waterblasts", are known to have been used at Lakeland mines. The best example was perhaps the one used at Greenside Mine when the long Lucy Tongue Level was driven.

Water was taken into a pipeline high up the steep gill and led down the mountainside to the baffle box at the level mouth; the resulting air was conveyed in pipes fixed along the side of the tunnel and reached well over half a mile. The devices were used at the mines on Caldbeck Fells where long adit levels are common. When the 90 Fathoms Level at Driggith Mine was reopened after being closed for 100 years, the remains of a baffle box were found at the foot of No. 1 shaft, down which water had been led in a 4″ square wooden box from the 75 Fathoms level.

The air pipes used had been only 5″ diameter and were made from pure zinc sheeting which the acid in the water had almost completely eaten away. The joints had been wrapped with strips of sacking soaked in Stockholm tar to prevent leakage.

The Deep Level at Tilberthwaite was ventilated by a waterblast and has been driven the 1,000 yards from the level mouth to the main copper workings without any intermediate shaft. When the eastern part of the 90 Fathoms Level at Roughtongill mine was being driven the company sank a small shaft by the side of Blea Gill, which is fully a quarter mile from the last ventilation point in the level. The level was draining off a large quantity of water when it was still some distance away but it failed to drain the shaft; there was difficulty in working before the connection was made.

In the old records of Driggith Mine boys are mentioned as working fans. These were small rotary type blowing fans used to bring some fresh air to dead ends and it must have been a miserable task for a boy set to work one of these machines day after day. Latterly, at Greenside and in the Caldbeck Fells barytes mines, compressed air-driven fans were used to either blow or suck through 12″ diameter steel tubes. At Force Crag Mine today the adit level is ventilated by one of these fans blowing fresh air into the face (which is now 900 feet away) through 12″ diameter heavy plastic tubes; they are both leakproof and fireproof.

Drainage of mines is a problem as old as mining itself and has always been a costly item however it was accomplished. Some of Lakelands mines have contained large springs which were easily dealt with where adit levels were possible, as on Caldbeck Fells, where very little mining took place below the deepest adits. Almost every mine has been partly drained in this way and every effort was made to catch the surface water at the adit so that the pumps had only the actual springs, or "make", with which to cope.

In the Coniston Mines the sole of the Deep or Horse Level was very carefully waterproofed in any parts where the copper veins had been stoped away below it. In these places a row of strong cross timbers was set some 3 or 4 feet below the actual level sole and after covering them with poles placed lengthwise (and on these a layer of flat stones) the rest of the working was filled up to the level bottom with well-trodden fine boulder clay. The considerable quarry where this material was obtained can be seen outside the Deep Level mouth and so well done was this work that after the pump was stopped it took the Bonsor Mine 5 years to fill to the adit.

At Greenside there were numerous launders in the higher old levels which carried streams of water over old stopes. When I was a boy I accompanied the old foreman on regular inspection trips and so was shown into long-forgotten parts of the mine. Greenside was reckoned a dry mine because the make of water from the great depth reached did not exceed 50 gallons per minute.

In contrast the Caldbeck Fells mines made large quantities of water which generally found its way down the porous vein material to the lowest adit. At Potts Gill barytes mines the flow from the Endeavour and Tunnel levels exceeded 200 gallons per minute even in the driest weather, and the various levels at Sandbeds Mines also discharged a total of not less than a similar amount. At Roughtongill Mine the flow from the extensive 90 Fathoms Level workings was taken over many years ago by Carlisle Corporation for a water supply and never again can there be any mining within the surrounding watershed. The lead mines in the Keswick area all had trouble with water and, as most are shaft mines, pumping was a problem.

The favourite pump used in the 19th century was the Cornish type in which the actual plunger was situated near the bottom of the shaft and operated by means of a string of heavy wooden rods lifted up and down in the shaft by either a waterwheel, steam beam engine or sometimes an hydraulic engine.

On Caldbeck Fells these pumps had been used at Haygill and Carrock End Mines, where they had been driven by waterwheels, and in Roughtongill and Driggith, hydraulic engines supplied the power. Thornthwaite Mine, 500 feet deep, was drained by one of these pumps worked by a Bull hydraulic engine. At Brandlehow where the flow was 150 gallons of water a minute, steam power was used for over 40 years and the cost of coals might have been avoided had a long water leat been constructed to bring water from the Derwent with which to turn a large waterwheel.

When the deeper parts of Threlkeld Mine were being worked it was found that the water was so acidic that it soon ate away the valves and their seats in the pumps and only those made of brass lasted any length of time. Near the far end of No. 1 Level at Force Crag Mine there is a spring in the floor which was discovered about 1855 and has since given off 100 gallons per minute of slightly warm water accompanied by carbon dioxide gas. Of all the Lakeland mines only the Myers Head Mine at Hartsop was actually lost because there was so much water the pump was overwhelmed.

The pumps in Greenside Mine were of centrifugal type direct, connected to electric motors, and the 1400 feet of depth below the adit was overcome in three stages. The deepest single-lift pump in the district was at Bonsor Mine where the largest waterwheel in England worked a Cornish type pump which delivered the water from the 170 fathoms level to the adit. The strangest pumping and winding arrangement was at Roughton Gill Mine where after much trouble with a water turbine at an underground shaft it was decided to install a portable type of steam engine. How it was ever taken through the long twisted levels to this remote spot is a mystery but when it was assembled it is said to have worked very well and enabled a great deal of ore to be won from below the adit. Local history says that the engine and its boiler were left in when the mine closed in 1878 but it has long been impossible to enter these workings to find out.

The pumping plant at Carrock End mine must have looked quite impressive, for the waterwheel was about 200 yards away from the shaft and was connected to it by means of horizontal rods which crossed under the main road. The water for turning the wheel was brought along a leat from Carrock beck and was first used for washing the ore near the site of the shaft then being discharged into the Black or Dutchmans Moss to flow across the road and onto the pumpwheel.

When the Company working the mine decided to erect the pumping plant they brought a suit against the Driggith Mining Company for loss of the water taken by them to Sandbeds by leats from Carrock beck and discharged by them into How beck. The Driggith Company had to construct a return leat to bring the water back from Sandbeds to Carrock beck where it entered some way upstream from the Carrock End mine leat. An interesting point is that after use at Carrock End the water flowed down the Black beck to the river Caldew. It would in any case have been quite impossible to return it to Carrock beck.

Transporting the ore underground.

WHEELBARROWS were employed in the mines for centuries and even after rail tracks and mine cars were in general use the wheelbarrow moved the broken material over short distances. The old men's barrows were constructed without legs and had large diameter cast iron wheels which made for easier wheeling over rough ground. When long distances were to be covered slings which went around the neck were fitted to the barrow shafts and so took much of the weight off the wheeler's arms. Leather flaps fixed to the outside of the shafts protected the hands from being skinned on the sides of the levels, which were generally very narrow and twisted.

Even in mines where rails were laid in the main levels it was usual to use wheelbarrows in the crosscuts and side drifts and the stuff from them was wheeled out to the main level and then reloaded into the mine cars. In some of the old levels ruts have been worn in to the solid rock by the passage of innumerable wheelbarrow loads. In the barytes mines modern rubber-tyred wheel barrows were used in short sub-level workings where it was not deemed worthwhile to haul up rails and a "bogie".

One can only guess at what date tram rails and "bogies" or mine cars were first used in the Lakeland mines. Wooden rails date back to at least the 17th century. Some of these were found stacked in the side of one of the Driggith levels and were similar to those seen in old workings at Patterdale and Dale Head. They were about 3 inches square and made of elm and were pegged down to the sleepers which were generally

A locomotive in the 90 fms Level at Driggith Mine.

of oak. After being laid, a 1″ wide by $\frac{1}{4}$″ thick iron strap was nailed along the top with its edge flush with the inside of the wooden rail so that it prevented the rail wearing and gave the flanges of the wheels of the cars something to bite against. All the nail holes were countersunk and special nails were used so that the heads did not protrude. No curves were seen in the Driggith workings but some were a feature of the Hartsop lead mine and had undoubtedly been made in the joiners shop. The track was carefully jointed and pegged down onto the oak sleepers so the joiner was possibly responsible for putting it in.

The first bogie wheels were of cast iron and large diameter and usually had a square centre hole for fixing to a square-ended axle by means of hard wood and iron wedges. Iron became more plentiful and so wrought iron rails came into being, known first as "razorbacks" for they had only a very small head and no base and were fixed in cast iron chairs by means of wooden or iron wedges. Several of these chairs and an odd length of this type of old rail were found in the Driggith 60 fathoms Level.

The wrought iron bridge-type rails could be nailed direct on to the sleepers and became popular. Their greatest drawback was the difficulty in bending them and some standard curves were supplied by the makers.

29

A crude form of bender was found in Driggith mine; it had been made in the smithy but it is doubtful if it was much of a sucess.

Towards the end of the 19th century the flat-bottom type of rails were on the market, supplied in various sections and weights from 10 or 12 lb. per yard (for use with small hand trammed bogies or cars) up to 30 lb. per yard (for the very heavy haulage of long trains by electric locomotive such as was used in the Lucy Tongue Level at Greenside Mine). Twenty pounds per yard rails have been a popular size at most mines; when carefully laid with short distances between the wooden sleepers and fish-plated together this kind of track has given little trouble, used in conjunction with ponies or small battery-type electric locomotives to haul the trains of mine cars. In the long Horse Level at Threlkeld Mine mules were used to draw the trains but this is the only Lakeland mine where they have been employed.

The use of small horses and ponies in the long adit levels goes back to at least the early part of the 19th century for "horse levels" are frequently mentioned in the old records. Several of these levels reached over a mile into the fells as at Greenside, Threlkeld and Coniston Mines. On Caldbeck Fells, ponies were also employed in Roughtongill and Driggith mines to haul the ore and waste through the long adit levels. Over the years there must have been several generations of ponies and drivers.

In most mines ponies were succeeded by the introduction of small electric locomotives.

DRESSING THE ORE.

Grating.

WHEN the crude ore was brought out of the mine it was sorted to remove the pure pieces of the mineral and to discard as much of the waste material as possible. From very ancient times up to the end of the 1914-18 war the crude ore was tipped, one car at a time, into the back of small semicircular stone-built bins or hoppers which were called "teams". The bottom of these sloped gently towards the open front and was paved with large hard cobbles and in the opening was set a grate some 2 feet by 3 feet, with the bars running lengthwise. The spacing of the bars depended on the type of crushing mill installed; the grate was laid at the same slope as the bottom of the "team". Beneath this grate was constructed a walled drainlike passage into which fell all material passing through the grate and from where it could be shovelled out and wheeled away for the next process.

As each car or bogie load was tipped into the team, a boy washed the ore with water under pressure from a hose and picked out the obvious pieces of clean waste rock or "dead", and any pieces of solid mineral; he then raked some of the rest of the ore forward over the grate, again washed it and repeated the process, breaking up any large mixed pieces called "knockings" with a double-hand hammer, until the team was emptied ready to receive another car load. The pieces of solid mineral were put into a measuring box holding about half a hundred-weight and when full this was taken to the Ore House, tallied and emptied. Sometimes a second grade was made which was not so pure but was still good enough to smelt without further treatment.

In the early days when only stamps were available for crushing ores the spaces in the grate were only 1″ wide and the knockings were broken to about 2″ size, sorted again and raked over the grate to remove any fines before being wheeled away to the stamps. This then was the grating layout in Elizabethan times. Then two important changes took place. Firstly crushing rollers were invented and were usually set up in sets of three pairs. The first pair of fluted type was direct driven off the waterwheel shaft and so was very powerful. When these machines were available the size of the grate was increased to as much as 4″ between the bars and obviously a great deal more material could be dealt with in each team. The knockings were also broken to pass through the grate.

The next invention was the stone-breaker and when one of these machines had been installed the knockings were no longer broken by hand except so that they could be fed to the crusher.

Going back to the Elizabethan "grates", as this sorting process was called, the water used for washing flowed out beneath the grate and carried with it a certain amount of fine material called "smiddam" which was caught in a little tank or pit made for the purpose. The settled material was shovelled out and wheeled away to the next process.

Bucking

In the days before crushing rollers were introduced all the "smalls" (material which fell through the grate) were passed to the "bucking" shed. Here mostly women were employed pounding up the ore to about $\frac{1}{2}$″ size or less. These woman worked at a long bench which was divided into small cubicles boxed in on three sides but open at the front and which had a thick iron top. The ore was pounded with a special type of hammer called a "bucking iron" or "buckler". These hammers had one large flat though slightly curved face. The heaviest kind had a wooden shaft and the smaller ones a short iron strap-like loop handle which was grasped in the hand. The pounded ore along with the smiddam was passed to the next process, sieving.

Sieving.

This operation was the forerunner of jigging. The workman filled a round riddle or sieve with the crushed ore and, holding it by handles fitted on the sides, agitated it up and down in a tank filled with water. Presently the material sorted itself out into layers by specific gravity with the lightest material, the "tailings", mostly stone and quartz, on top, followed by a layer of pieces of mixed stone and mineral called "middlings" and finally, next to the sieve, a layer of clean mineral which was called the "head" and which was ready for the smelter. The fine material was washed through the sieve into the tank and this was periodically drained through a bunghole in the side and then cleaned out.

MINING IN THE LAKE COUNTIES

Buddling.

Meanwhile the broken knockings were being pounded by the stamps and the resulting pulp was being washed in long box buddles so arranged that the feed could be readily switched from one to another, enabling them to be cleaned out without stopping the stamps.

The fine material from the sieve tanks was also treated in a box buddle but this was done by men with long-handled shovels who worked the ore towards the head of the box against a little stream of water so arranged as to flow steadily full width of the bottom. This process was called trunking and the two men who stood one each side of the buddle were called "platers". A good deal of skill was required to obtain best results. They produced a head or concentrate, a middling and a tail.

These middlings and those from the sievers were mixed with the knockings so that the stamps were not supplied with all rough or all fine material and so were made to deliver pulp at maximum capacity.

The Elizabethans built mills of this type at Goldscope, Grasmere and Roughtongill Mines. When the crushing rollers were invented is obscure but they do not seem to have been in general use until the late 18th century. Sieving went out with the introduction of the hand jig about 1600 and before long a mechanised form of this machine was invented and greatly used at most of the old mines which had dressing plants and especially at Coniston.

Jigging.

These machines consisted mainly of a long box-like tank which could be drained of its contents by means of valves set in the front near to the bottom. A drive shaft with several cranks ran along the top mounted on timber frames. At each crank there was a sieve box about 2 feet wide and 3 or 4 feet long which was suspended from the crankshaft and which could be lifted up clear of the cranks by means of a lever and so put out of motion at will. A hopper was set behind the machine with spouts opposite each sieve box so that they could be readily filled with crushed ore. One operator looked after a batch of sieve boxes and was kept busy emptying them though it was a much lighter task than that of the old time "sievers".

Modern jigging machines came into general use about the middle of the 19th century and worked somewhat on the same principle but were continuous in operation and did not require the sieves to be constantly cleaned.

Stamping Mills were at work at the Triddle Mine at Coniston and also at Roughtongill, Greenside and Driggith Mines in the 18th century. Mills with crushing rollers were built at Driggith in 1812, Greenside about 1830 and at about the same time at Bonsor and Paddy End Mines at Coniston. Eventually in the 19th century there were 13 dressing plants incorporating crushing rollers at work in the district as well as 2 stamping mills.

The box buddle was replaced by Zennor's rotary buddle, first erected at Driggith Mine by Capt. William Jeffereys in 1858. It was so successful that soon every other plant installed them.

The roller mill at Driggith did very well at first but there seems to have been trouble through the very hard quartz tending to wear gutters in the chilled iron rollers for manganese steel ones did not come until this century. These gutters let more and more partly crushed material pass and eventually a heavy overload of oversize pieces built up thus greatly decreasing the efficiency of the plant. In 1822, when it was decided to erect a new mill at the foot of the hill below the Sandbeds Level (60 fathoms level), stamps were installed instead of rollers and these were worked continuously for 50 years. As most of this ore is extremely hard the losses due to the galena being slimed must have been considerable and it is doubtful if more than 70% was recovered.

When Bonsor mine was in full production in the 1840s there was a large collection of Cornish type stamps which are said to have been heard in the village night and day except on Sundays.

The all gravity-type dressing plant at Greenside Mine in 1929 was able to make a plus 90% recovery of the galena in the crude ore.

The late Anthony Wilson of Thornthwaite, perhaps the most expert 20th century ore-dressing engineer in Lakeland, erected gravity-type plants at Carrock, Threlkeld and Thornthwaite mines to recover tungsten at the former and galena and blende at the others. Carrock Mill was the best gravity-type plant ever built in Lakeland for it incorporated the then very latest concentrating tables for treating sands and slimes, cutting losses of the very fragile tungsten minerals to a minimum.

One of the very first floatation type mills was erected at Coniston. There was also a roasting, leaching and electrolytic plant which produced sheet copper 99.9% pure direct from the ore in 1912.

A floatation plant of Elmore type was erected during the 1914-18 war at Force Crag mine to separate barytes and blende from a mixed concentrate.

In 1936 the new mill constructed at Greenside mine was able to treat 250 tons of crude ore in 24 hours and combined both gravity and floatation thus making the highest possible recovery and, at the same time, producing a high grade lead concentrate.

In all the years since the coming of the Elizabethan Dutchmen a total of 38 ore dressing mills have been erected in Lakeland and not one remains today. The following is a list giving location, date, type, prime mover and minerals obtained.

Mine	Type
Roughtongill	Stamps & box buddles
Goldscope	Do.
Grasmere	Do.
Driggith	Do.
Greenside	Do. and possibly hand jigs
Triddle	Stamps & box buddles
Driggith	Crushing rollers, jigs, box buddles
Driggith	Stamps with Zennors buddles
Bonsor	Crushing rollers, stamps, jigs, buddles
Paddy End	Do.
Greenside	Do.
Greenburn	Crushing rollers, jigs, buddles
Hartsop	Do.
Redgill	Stamps, box buddles
Helvellyn	Crusher, rollers, jigs, buddles
Goldscope	Crushing rollers, jigs, buddles
Force Crag	Do.
Thornthwaite	Do.
Brandlehow	Do.
Roughtongill	Crusher, rollers, jigs, buddles
Greenside	Crusher, rollers, stamps, jigs, buddles
Tilberthwaite	Crushing rollers, jigs, buddles, stamps
Threlkeld	Crushing rollers, jigs, buddles
Barrow	Do.
Potts Gill	Crushing rollers, jigs
Wetherlam	Crusher, stamps, table
Carrock	Crusher, rollers, jigs, tables, buddles
Force Crag	Crusher, rollers, jigs, tables, ballmill
Driggith	Crusher, rollers, jigs, buddles
Coniston	Crusher, ballmill, floatation plant
Potts Gill	Crusher & jigs. at Caldbeck
Potts Gill	Do. at mine
Hartsop	Crusher, rollers, jig, tables, stamps
Greenside	Crusher, ballmill, jigs, tables, floatation
Potts Gill	Crusher, rollers, jigs, tables
Force Crag	Do.
Driggith	Crusher, rollers, jigs
Force Crag	Crusher, rollers, jigs, table

Date	Power	Minerals
1569	Waterwheel	Lead & Copper.
1569	Do.	Copper.
1569	Do.	Lead.
1700	Do.	Lead.
1700	Do.	Lead.
1750 about	Do.	Copper.
1812	Do.	Lead.
1822	Do.	Lead.
1830 about	Do.	Copper.
Do.	Do.	Copper.
Do.	Do.	Lead.
1840 about	Do.	Copper.
1850 about	Do.	Lead.
1861	Do.	Lead & Copper.
1861	Water turbine	Lead.
1850	Waterwheel	Lead & Copper.
1830	Do.	Lead & Zinc.
1848	Do.	Lead & Zinc.
1848	Do.	Lead.
1866	Do.	Lead & Copper.
1871	Do. & turbine	Lead.
1871	Waterwheel	Copper.
1879	Do.	Lead & Zinc.
1883	Do.	Lead.
1888	Do.	Barytes.
1903	Oil Engine	Copper.
1905	Water turbine	Tungsten.
1905	Do.	Lead & Zinc.
1906	Steam Engine	Lead & Zinc.
1912	Waterwheel & turbine	Copper.
1915	Waterwheel	Barytes.
1926	Oil engine	Barytes.
1931	Do.	Lead.
1936	Hydro Electric	Lead.
1938	Oil engine	Barytes.
1940	Do.	Barytes.
1948	Diesel engine	Barytes.
1961	Diesel electric	Barytes.

The Caldbeck Fells

CALDBECK Fells were a source of minerals even before Elizabethan times and each century saw a prosperous revival of mining. The Elizabethan miners, who were mostly Germans, re-opened several old workings, the chief of which was the famous Roughtongill Mine, and here they erected one of the very first crushing and dressing plants in Lakeland.

Caldbeck village was well-established even then, and eventually it contained not only woollen and blanket mills but a brewery and a paper works. A large bobbin mill was erected in the Howk but this was not until the 19th century. Every country trade was represented, even a nail maker. The old saying "Caldbeck and the Caldbeck Fells are worth all England else" must have had real foundation in the great prosperity of those far-off days.

Whether the great work at Roughtongill came to an end with the destruction of the smelter at Keswick by Cromwell's forces in 1651 is not known and there are no records for the rest of that century. No doubt some mining would be going on, if only on a small scale, but the nearest lead smelter was probably at Alston and carriage of the sorted ore would be a costly item. At the beginning of the 18th century the rich outcrop of Driggith was discovered; it is strange that the Elizabethans had failed to find it, especially as the lode is the same as the one they were working at Roughtongill.

This find put new life into the area because the proprietors erected a smelter near Haltcliff and so the problem of carting the ore long distances to be smelted disappeared. Towards the end of the century a smelter is said to have also been at work at Roughtongill but there are no output figures for any of the mines during this period; it seems certain that they were in more or less continuous operation.

The beginning of the 19th century saw a great deal of mining activity similar to that experienced all over England for, with the growth of industry and the onset of the Napoleonic wars, more and more lead and copper were required and the greatest search of all time for new deposits was started. Roughtongill and Driggith worked steadily and produced a considerable tonnage of lead, copper and silver, although by far the greatest part of the output was before any official records were kept and unfortunately few of the estate records or plans of the mines have survived. By 1874, Driggith was closed through a variety of reasons and the great Roughtongill came to an end in 1878, to all intents and purposes exhausted.

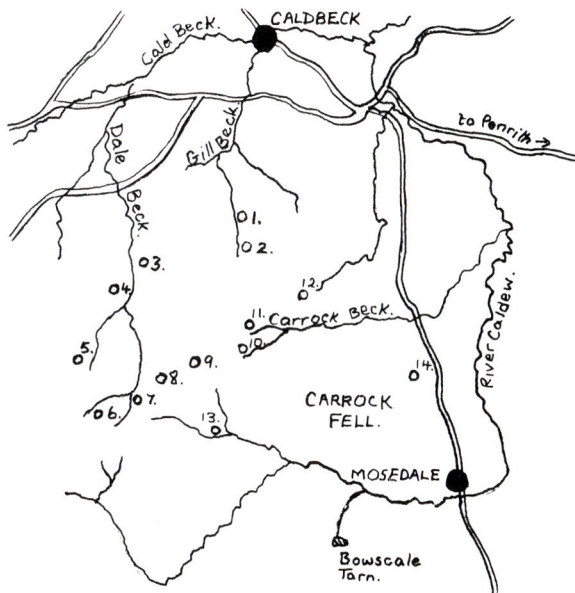

The major mines of the Caldbeck Fells.

KEY
1. Potts Gill Copper Mine.
2. Potts Gill Mine — Barytes.
3. Haygill Mine. — Copper.
4. Brae Fell Mine. — Lead.
5. Redgill Mine. — Lead.
6. Silvergill Mine. — Lead.
7. Roughtongill Mine. — Lead.
8. Mexico Mine. — Lead.
9. China Clay Mine. — Clay.
10. Drygill Mine. — Lead.
11. Driggith Mine. — Lead, Copper.
12. Sandbed Mine. — Barytes.
13. Carrock Mine. — Lead.
14. Queen Mine. — Copper.

Many small syndicates tried to revive the field but without much success and there was little new work done until in the early 1900s when operations were begun in earnest at the Carrock Wolfram Mine. This was worked on a considerable scale until the end of the 1914-18 war, when the price of wolfram fell away and the mine became un-economic. About the same time the barytes deposits at Potts Gill attracted attention and eventually were opened out to give a continuous operation on a profitable basis for over 25 years. For a long time they have been the only mines in operation on the fells.

Such then is the outline history of the field which is still a paradise for the mineralogist and geologist, although the chances of finding good mineral specimens, especially the rarer varieties, are not now as good as formerly. Most of the mines have been closed for nearly 100 years and the heaps have been gone through over and over again.

Briefly, the Caldbeck fells occupy the northern part of the Skiddaw Fells and cover an area stretching southwards from Caldbeck to the river Caldew in the Mosedale valley. Westwards, they run to the eastern flanks of Knott and Brae Fell and the eastern boundary is the Hesket Newmarket - Mungrisedale road. The chief peaks are: High Pike, 2157 feet; Great Lingy, 2009 feet; Knott, 2329 feet; Brae Fell, 1920 feet and Carrock Fell, 2174 feet.

Apart from Carrock Fell, there is very little bare rock to be seen as most of the area is covered in thick boulder clay with some peat. Deep gills are numerous and it is in these that most of the lodes have been discovered. Some hushing was done by the ancients and good examples are still to be seen at Brae Fell and Carrock End. Supplies of water for dressing and for propelling the water wheels, driving the crush-ing and dressing machinery, were very important and miles of old leats can still be traced around the hillsides.

Many of the springs issuing from the old workings are of considerable volume and some of them have now been taken over for water supplies by the Local Authorities. Roughtongill, Hay Gill and, to a lesser extent, Sandbeds, all supply water and this has sterilised these areas for future mining. No doubt the springs at Potts Gill Mine will someday also provide much needed water for the built-up districts to the north. It thus appears inevitable that all mining will eventually come to an end as it will be impossible to drive any long prospect tunnels without fear of draining one or other of these water supplies.

The geology of the area is complex as can be seen from a map. The variety of the minerals obtained is, no doubt, related to the diversity of the rocks encountered. By far the greater part of the fells is composed of the two lower members of the Borrowdale Volcanics—the Higher Ireby Group of tuffs and andesites and the Eycott Hill Lavas. The Carrock Fell gabbro granophyre mass, bounded by the Skiddaw granite to the south of Carrock Mine, reaches up to Roughtongill North Lode and westwards the Skiddaw Slate comes in. In Dry Gill there is an isolated patch of caradocian shale containing fossils. The whole area is criss-crossed by lodes and crosscourses and it is readily apparent that there have been many periods of movement and mineral deposition.

THE CALDBECK FELLS

Much of the Borrowdale rock has been oxidised to varying depth below the present surface. Not all of this decomposition can be attributed to weathering for there is little doubt that acid solutions have had a great deal to do with it, particularly where traversed by lodes. This altered rock has, so far, been found to carry almost all the profitable mineral ground for lead, copper and barytes and where the unaltered rock is reached the lodes become filled with stone, quartz and carbonates.

No major trial has been made to test them at greater depth to see if there is a lower zone of enrichment. At Roughtongill little or nothing of importance was seen in the lowest shaft workings and at Potts Gill the lodes are very restricted and almost completely barren in the lowest adits. The latest geological movement is the NNE crosscourses which are filled with "douk", a clay-like substance containing no mineral of any sort. These crosscourses were preceded by the WNW barytes-bearing lodes which are shifted by the crosscourses but pass straight through the rest of the lodes. It is thought that the east-west quartz lodes, which carry a little lead and barytes, are the next youngest, followed by the NE-SW lead-zinc lodes. There are also north-south quartz lodes which carry only slight mineralization, NW quartz lodes which carry lead and copper and last of all east-west copper lodes. Almost all of the lodes carry manganese minerals in the upper parts and this is sometimes the chief constituent. The manganese occurs both as soot or wad and psilomelane with pyrolusite occasionally.

As far as can be seen, barytes occurs in almost all the lodes but, apart from an area on the Driggith Lode, the workable ore is confined to the WNW lodes of Caldbeck Mines area (embracing Potts Gill and Sandbeds sections). Carbonates occur in most of the lodes but especially so in the WNW barytes lodes and the NE-NW lead-zinc lodes. Quartz in varying degrees of hardness is the chief gangue mineral in all the lodes.

The most disappointing feature about the mineralization on Caldbeck Fells is the apparent failure at very shallow depths compared with the deep-seated ore shoots of Patterdale and Coniston, which also occur in Borrowdale Rocks but which rocks are virtually unaltered, even on the highest outcrop

The failure of the ore in depth at Roughtongill was for a different reason as the workings were not in the Borrowdale Volcanics and have no relationship to the Patterdale and Coniston ore bodies.

There does not appear to be any other old workings on the Fells worthy of re-opening and the chances of any large orebody having been overlooked by the ancients is extremely doubtful. In all the years between the 12th century and the present day many hundreds of men must have prospected the area and, judging by some of the veins they discovered, they appear to have virtually left no stone unturned. Modern science has not so far produced any instrument which can locate small ore-shoots beneath glacial till and so we are not much cleverer than our forefathers in this respect. Indeed, few today have the know-how and nose for ore that apparently they had.

Thus, almost 100 years after the closure of the lead mines, due largely to the exhaustion of the worthwhile ore, the barytes mines close for mainly the same reason. In the last 27 years the Caldbeck barytes mines have been intensively worked and have on the whole been a profitable operation as well as providing traditional work for the local men.

Roughtongill Mines.

THE lead and copper mines of Roughtongill, now long abandoned, lie at the head of Dale Beck, some $3\frac{1}{2}$ miles south-west of Caldbeck, and are extremely ancient and extensive. They are thought to have been started in the 12th century but were idle when taken up by the Elizabethans in 1566. They were worked quite vigorously by them for several years.

Little is known of the next 200 years but in 1710 Lord Wharton was working the Silver Gill section of the mines, apparently with some success. Whether these operations carried on continuously is not known although the mines were very busy towards the end of the 18th century, the smelter being built in 1794. From then on the mines were worked enthusiastically until 1878 when they finally closed, to all intents and purposes worked out.

Two main lodes coursing NE-SW and heading NW have been extensively worked, although the more southerly or Great South Lode has been by far the richer of the two. Other lodes worked were Caunter, Dobsons and Silver Gill, but none of them extensively. The geology of the area is rather complex and the variety of the wall rock plays an important part in the productiveness of the lodes.

The Great South Lode, running NE-SW and dipping NW at 20° is, at Roughtongill, bounded on the footwall or south by granophyre from Crown Point to near the junction of the two Thief Gills. At this point there is a patch of gabbro followed westwards by a small area of granophyre and then continuous Skiddaw Slate.

The hangingwall or north side has Borrowdale Rocks at Crown Point followed by granophyre at Mexico Mine. Gabbro comes in halfway along the front of Iron Crag and continues throughout Balliway Rigg. At the head of Silver Gill is a patch of decomposed fellsite followed eventually by Skiddaw Slate. The best orebody occurred where there was granophyre on the footwall and gabbro on the hangingwall. Over much of the rest of the ground the lode was more or less barren.

The North Lode is roughly parallel to the South Lode and also heads NW, but at a steeper angle. To the east it has Borrowdale Rocks on both walls and is virtually barren. Where the gabbro is on the footwall it carries ore-shoots which, although quite payable in parts, are nothing like as rich as those on the South Lode. Connecting the South and North Lodes in the gabbro is the Caunter Vein which runs diagonally. The ore at the junction with the North Lode on the 90 fm. Level was very rich but died out completely on the level 20 fathoms below.

After the failure of the New Engine Shaft to find any ore at the 20 fm. level, the company decided to abandon this project and to sink an underground shaft at the junction of the Caunter and North Lodes. At first good progress was made but the water flow became too heavy for hand haulage and water was brought down in pipes from the 60 fm.

SCALE :— 60 FT. TO 1 INCH.

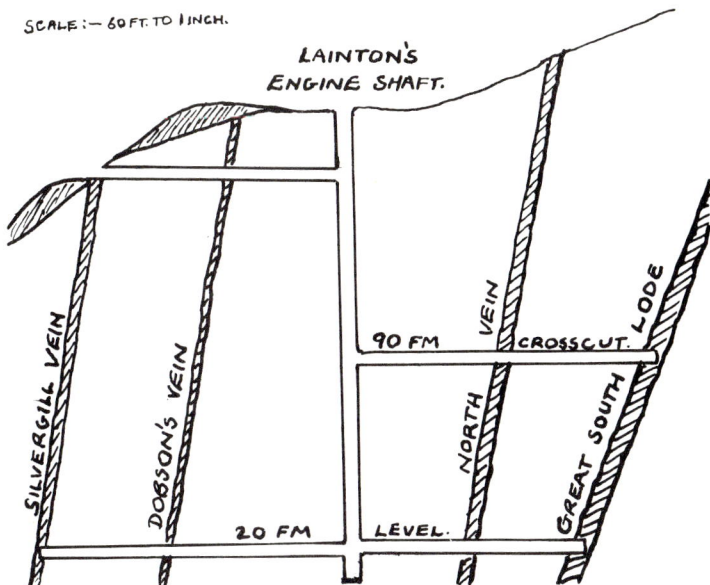

LAINTON'S
ENGINE SHAFT.

SILVERGILL VEIN

DOBSON'S VEIN

90 FM

VEIN

CROSSCUT.

GREAT SOUTH LODE

NORTH

20 FM

LEVEL.

A cross section of the Roughtongill Mine.

Level to work a turbine for winding and pumping. This machine gave nothing but trouble and the shaft was more often flooded than workable, a state of affairs which continued for two years.

It was decided to shoot out a place at the shaft top to install a 30 feet diameter by $2\frac{1}{2}$ feet breast water wheel, and some very good copper ore was found through the hard spar called the "fellspar" which is three to eight feet wide on the north side of the lode. The water wheel project was given up in favour of a small portable steam engine. It worked quite well and all the worthwhile ore above the 20 fm. Level was removed. This 20 fm. Level worked both the North and South Lodes and in September, 1875, the company were sinking below the level on the South Lode in good ore. By the end of October the ore disappeared without trace and was never again found.

Northwards is Dobson's vein, which has yielded a little ore, and this is followed at a further short distance North by Silver Gill Vein. Both these lodes are entirely in the Borrowdales and run into the North Lode to the NE of Dale Beck.

The Skiddaw Slate comes in at the head of Silver Gill and there is evidence of faulting along the breast of Yard Steel. To the west of the head of the gill the continuance of both these lodes is doubtful although it would appear well worthy of a trial as there is good lead ore, both carbonate and galena, in the tip of an old shaft at the head of the gill.

MINING IN THE LAKE COUNTIES

Why the mineralisation failed at a shallow depth below the 90 fm. Level in every case is difficult to ascertain. A letter written by William Muse, Lord's agent, to Sir G. W. Denys dated 26th October, 1875, stated that the ore below the 20 fm. Level was entirely cut off and no trace of it could be seen. This suggests faulting but there is no proof of this and it may have been due to a change in wall rocks. It is presumed that the case mentioned in the letter was the Junction Shaft workings on the Caunter and North Lodes but apparently the South Lode was also worked to the 20 fm. Level and so it is not clear which working is meant.

The smelter, built in 1794, was operating in 1853, and capable of smelting 100 tons of lead per month. Coal was at that time delivered to the smelter for 12/– per ton. By the time the Caldbeck Consolidated Mining Company Limited came into being (1865) the smelter was disused and was made into a row of cottages. The lead and copper ore was carted to the railway at Wigton.

This Company built a new dressing plant at Roughtongill which was capable of making a much better separation of the sometimes complex ore and concentrates of galena or blue lead, carbonate or coloured lead, copper ores and zinc blende were sold. The South Lode was opened at the outcrop on the north side of Thief Gill by several small pits which are probably the earliest workings as they are certainly pre-gunpowder.

Professor W. G. Collingwood translated the names of the various Elizabethan workings from the Augsberg records of the Company of Mines Royal but there is now no way of knowing which they were.

In 1569 there was the Lady on Horseback, Lower Nick, St. Emanuel, New Adit, St. David, Luck, St. Elizabeth and St. Mark, but it is doubtful whether these were all at Roughtongill proper.

A stamp mill with a house for the men was constructed mostly from timber obtained in Parson's Park at Caldbeck, the iron work for the water wheel and the stamps being brought from Keswick. Wolff Hochholzer took a contract to build this Stamp Mill for £20 but ran into so much difficulty that he lost money. The company held an inquiry in to the matter and three carpenters on oath before the Company's Chief Officials judged that £36.12s.6d. was a just price for the job and this sum was agreed. Hochholzer completed the contract for this price but his own completed figure was £46.

There were 79 people employed at the mines in 1571. This labour force was made up of the mine foreman, stamp foreman, carpenter and 23 miners (all Germans), 5 English sorters, 11 English carpenters, 33 English labourers (some in the mine, some outside), 3 English carriers, and one English woman who looked after the house. How long this activity continued is not known but it must have come to an end when the smelter at Keswick was destroyed in 1651, for which act Cromwell's army is generally blamed.

The date of the 30 fm. Level, which is a crosscut from beside the gill, is not known but it is believed to be 18th century and was the main working until about 1830 when the 60 fm. Level crosscut was driven and in 1849 the 90 fm. Level was commenced.

The 30 fm. Level crosscut starts just south of the outcrop of the North Lode and has been driven SW along the South Lode for over 300 fathoms. In 1865 this face was still being driven and after going through 120 fathoms' length of barren ground came again into ore which was very mixed lead, zinc and iron pyrites and letting out a feed of water. By 1867 they had reached the end of this run of ore and after crosscutting in extremely hard rock cut a vein of good ore 3 feet wide. As there is no further mention of this face in the old records there is no idea of the state of the vein in the present end of the level. Eastwards, the level was driven along the vein towards Iron Crag and much ground stoped out nearly to the surface.

In 1822 the mines were being worked by a Company of local miners who sold Mr. J. Dickinson of Alston a quarter share in 1830. With this new capital the 60 fm. Crosscut was driven 70 fathoms to the South Lode, cutting the North Lode about a third of the way on. Over the years this level was extended SW along the South Lode for over 300 fathoms and opened into a great bunch of ore 70 fathoms in length. This bunch was 6 to 12 feet wide and often wider, composed of gossan, friable quartz, calcite with much phosphate and carbonate of lead on the fringes, but as the miners got further into the bunch the orey part was found to be mostly cubical galena carrying a more than average quantity of silver. By 1853 much of this ground had been worked out above the 60 fm. Level and had yielded large returns.

A rise was put through to surface in Thief Gill and water led down a pipe line to a Hydraulic Engine. By this means a shaft was sunk to the 70 fm. Level and the same run of ore was worked; it proved even richer.

The 90 fm. Level, begun in 1849, is a crosscut 120 fathoms in length driven S22°E to cut the South Lode. It also passes through the Silver Gill, Dobson's and North Lodes. South-westwards it has been driven over 350 fathoms along the South Lode and at 300 fathoms SW the lode is 30 feet wide for over 30 fathoms. Much of this ground was very difficult to work but yielded much excellent ore.

East of the crosscut the level was driven vigorously for many years and extends right up to the Mexico Mine. An air shaft was sunk by the side of Blea Gill and was connected with the level in 1867. Throughout the whole of this driving no workable ore was met with but the lode was extremely large and strong and almost always contained a little phosphate ore; this no doubt led the Mining Company to believe that deep down there would be rich primary deposits. An engine shaft (Lainton's) was therefore commenced in 1866, about halfway along the hillside between Thief and Blea Gills and sunk down towards the 90 fm. horizon. A crosscut was driven to it from that level on the South Lode and a connection made. The shaft was then sunk on to the 20 fm. Level. Two horizontal steam boilers, a Cornish pumping engine and a horse whim were erected and put to work. At the 20 fm. Level, crosscuts were driven north and south cutting the veins as follows; southwards —11fm. 5ft. to the North Lode and from there to the Great South Lode —11fm. 5ft.; northwards 20 fm. to Dobson's Vein and from there to the Silver Gill Vein—12fm. 5ft.

On the 60 fm. and the 90 fm. Levels, drifts were driven along the

North Lode with internal 30 fm. and 80 fm. Levels. South-westwards the 90 fm. Level was being driven up to the cessation of operations. The last available report, dated December 30th, 1875, states that the lode in the face was 10 feet wide and although containing no ore looked well. Six months previously it was in good ore and was in new ground ahead of the upper workings. Eastwards, the level was driven towards Mexico Mine and at times was in good ore but the total distance driven is unknown.

The Mexico Mine, which is part of Roughtongill, lies between Blea Gill and Clints Gill and the surface workings are old. The main workings are on the South Lode, which is very large and consists almost entirely of quartz. Most of the lead occurs as pyromorphite though there is a little galena. Copper also occurs, mostly as red and grey oxide, but only in small quantities.

Apart from the Shallow Level, 40 fm. and 60 fm. Levels have been driven and the 90 fm. Level has been connected to these workings. Apparently all this work proved abortive as the quantity of ore obtained in no way compensated for the very extensive development undertaken.

The workings in Silver Gill are also part of Roughtongill Mine and the highest level, known as the Golden Vugh, is extremely ancient. In 1853, a 60 fm. Level was being driven from the side of the gill, which is some 35 fathoms below an older, higher 20 fm. Level. The lode was 3 to 12 feet wide and composed of douk quartz and calcite with galena and a little copper. The ground was easy to drive and it was confidently expected that rich ore would be found beneath the higher ancient workings but there are no records of any being discovered or in fact whether the level was extended far enough. The 90 fm. Level was given up in 1863 in barren ground, but again it is not known how far it has been driven.

When the mine closed in 1878 most of the plant and buildings were removed and the cottages became derelict.The area was taken up again by the Cleator Moor Iron Company in 1888. They were looking for umber and barytes but nothing was done underground and this undertaking came to an end in 1894. Eventually the mines were taken over by Carlisle Corporation as a source of water and thus no further mining can take place.

According to Postlethwaite, no less than 23 distinct minerals have been recorded from these veins though the chances of finding any good specimens on the dumps are now remote, and all the workings have long been closed.

Red Gill Mine.

THESE lead workings lie about $\frac{1}{2}$ mile WNW of Roughtongill Mine and just west of the junction of Red Gill with Swinburn Gill. Some of them are reputed to be very ancient. From 1861 to 1866 there was a period of continuous operation but the output was very small indeed.

	lead concentrates	copper ore
1861	$1\frac{1}{4}$ tons	
1862	15 ,,	
1863	9 ,,	
1864		
1865		2 ton 8 cwt.
1866	$2\frac{1}{2}$,,	
	$27\frac{3}{4}$ tons	2 ton 8 cwt.

There are three lodes shown on the geological map but two of these are faults which bear ore only in part. The main lode courses NW-SE, dipping NE, and runs across the tongue of ground between Red Gill and Swinburn Gill. It appears to cut through the NNE-SSW fault vein which more or less follows the course of Swinburn Gill. Three crosscuts have been driven into the tongue with the lowest, No. 1 or Deep Level, from the side of the beck. The No. 2 or Old Dutch Level Crosscut is 16 fathoms higher and cuts the lode at 6 yards in. It continues some way further SW and at 8 yards from the lead lode a small copper vein has been cut. No. 3 or Top Level is 5 fathoms higher up and is very old, although some work was done here about 1870. A very ancient High Level, direct on the vein, is 23 fathoms higher up the gill but does not extend very far and has long ago fallen in.

Below No. 2 Level are several tons of material which shows a good deal of lead. A fair stream of water issues from No. 1 Level but all these workings have been closed at the entrances for many years.

The Top Level starts from a short crosscut SW and is driven NW for 50 fathoms with a rise to surface just SE of the crosscut. It is very likely

SE.   NW.

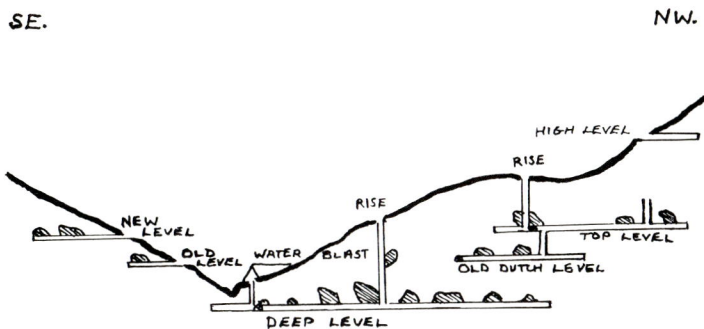

SCALE :— 50 FMS TO 1 INCH.

A longitudinal section of the Red Gill Mine.

45

that this level and the high Level are even older than the lower Old Dutch Level which is attributed to the Elizabethans. When the Top Level was re-opened about 1862 it was driven on for a short distance and a further small bunch of ore was discovered.

The Deep Level starts as a short SW crosscut from the side of the beck and cuts the vein about 5 fathoms below the gill bottom. A rise was put through at this point to enable a water blast to be installed for ventilation. The level extends 100 fathoms to the NW and has discovered eight short shoots or bunches of lead ore in this length. The total length of this ore is, however, only 30 fathoms and as the bunches do not extend very far either upwards or downwards it is obvious the mine did not pay and by 1866 these workings were closed. The company commenced a new deeper level 40 fathoms lower down the gill and continued on a small scale up to 1871.

Across the beck and on the west side of Yard Steel and east of the NNE fault are two levels directly on the lode, the upper one of which extends 16 fathoms and has found two little pockets of lead ore. These levels are also long ago fallen in but some nice pieces of lead ore can be found in the tips.

Higher up Swinburn Gill is the site of an old level at about 1750'OD. apparently driven on the course of the NNE-SSW fault. There is quite a stream of water from the old level and on the tip good pieces of galena can be found. All these workings are in Borrowdale Volcanics but southwards this fault vein will come into Skiddaw Slates and may be more productive.

Between this old level and No. 1 Level, beside the beck on the west side, is the site of an old level, the entrance to which has disappeared beneath a scree. The tip does not show any lead or copper ores but does contain some fair pieces of barytes. No sign of the lode can now be seen but it is totally different vein material to the Main Lode which is almost wholly quartz. It may be that it is a mineralised part of the NW-SE fault shown on the map, but if so then it is much further north than shown.

Red Gill Lode is famous for the very fine specimens of linarite which have come chiefly from the No. 2 Level workings. Other fairly common minerals found are: galena, chalcopyrite, malachite, cerussite and barytes, with rarer specimens of chrysocolla, leadhillite and caledonite.

A stamp mill was erected in 1861 and parts of it still remain. It was a Four Cornish-type stamp driven by a water wheel and it crushed the ore to coarse sand after which it was dressed in long box buddles.

Dry Gill Mine.

DRY GILL MINE, about $\frac{1}{4}$ mile SW of the old dressing floors of Driggith Mine, lies in a rough gill containing little water—hence the name. The occurrence is interesting in several ways, both geologically and mineralogically, but from a commercial standpoint it has not been of much value. The lode is thought to be the continuation of the Carrock End Copper Lode and is very wide, strong and composed almost entirely of quartz much of it extremely hard. The direction of the lode is east-west and the hade is northwards at 10° to 15° off vertical.

Much of the footwall rock is Carrodiocan shale known as Dry Gill Shale which occurs as an inlier with flinty altered ash to the south and weathered brown ash to the north. Three north-south faults, approximately 1,000 feet apart, cut through the lode and displace it to the left. These faults are thought to represent the barren continuations of the wolfram-bearing lodes at Carrock Mine. They seem to die out further north, however, as they have not been noticed to cross the Driggith Lead Vein. The most easterly fault continues almost up to this lead vein and has carried a little copper ore in the Driggith 30 fm. Level. Westwards, the Dry Gill Lode joins the Driggith Lode which is then called the Roughtongill South Lode. Strangely, the actual junction has not been tried.

Dry Gill is not thought to be a really old mine and possibly was opened about the same time as Driggith. Most of the lead occurs as mimetite or kampylite, the arsenate of lead which has a curved barrel-like crystaline structure. There is also some pyromorphite but galena is scarce. There is a little barytes and in places a good deal of manganese soot. A fairly rare oxide of lead and manganese, coronadite, is also found in small quantities.

The recorded output is small although the total area explored is also not large. In 1848-49 some 90 tons of concentrates were obtained. In 1850, however, only 2 tons were raised and there is then a gap until 1875 when $5\frac{3}{4}$ tons were obtained. It is said the coloured lead ores were sold to glass manufacturers. The largest working is a cross-cut level at about 1,750'OD., running north from the gill to the lode. It is from this level that most of the ore has been won and there is a stope through to surface which is still partly open. The lode here is in two parts and most of the workings appear to be on the northern branch. Midway between these levels a cross-cut has been driven south into the hill towards Brandy Gill. This is partly fallen and full of water but from the appearance of the tip it has opened into a vein which contains a little barytes and a good deal of quartz.

Carrock End or Queen Mine.

THIS old mine at Carrock End is the most easterly working on the Caldbeck Fells. It lies just west of the road from Mungrisdale to Hesket Newmarket, about a quarter mile north of Stone Ends Farm and at the foot of the steep slopes of Carrock Fell. It is not known when the mine was discovered but it was last worked by Captain Francis about 120 years ago. He is reputed to have made a fortune out of it.

The lode is on the faulted contact between the Carrock Fell granophyre and the Eycott Hill Lavas of the Borrowdale Series. It runs NW-SE at the mine and hades slightly to the east but westwards the contact curves round until it runs almost east-west at Dry Gill Beck, a distance of about two miles away. The faulted part containing the copper vein runs in the direction of Caldbeck but is barren after leaving the contact. The vein is mostly hard quartz with chalcopyrite and some malachite.

The lode is thought to have been discovered by the Elizabethans and had been worked prior to 1692 when a Dr. Edward Wright is said to have revived interest in the mine. In an old undated report the mine is referred to as at the Dutchmans Moss and there was then a Whim Shaft which appears to have been sunk from the opencut. The mine is referred to in *A Compendium of British Mining* by J. W. Watson of London in 1843. He says that in 1813 several thousand pounds worth of ore was raised but that although the sett was very extensive and contained several lodes it had never been worked to a profit.

Dr. F. H. Day in his paper *Some Notes on the Minerals of Caldbeck Fells* September 1927 (to the Carlisle Natural History Society) quotes from an old record dated 3rd October, 1794, as follows:

"A large copper vein had been discovered on the north side of Carrock Mountain. Trials had formerly been made at several places. It is five feet wide and the copper worth £30-£40 per ton. It is supposed that two workmen got £80 worth one afternoon last week. The present lessees are Wm. Rowe and Company of Liverpool. We have since heard that this vein has turned not out as well as was expected."

A hush made down the fellside cuts across the lode and there is a very old opencast at the site of the shaft. Probably these are the old trials mentioned above as the rest of the workings seem to be more recent and were probably made by Captain Francis. A vertical shaft, 23 fathoms deep, was sunk near the east end of the opencast, with levels off at 13 and 23 fathoms. The 13 fm. Level Crosscut is 8 fathoms long to the Main Vein and also cut two branches between it and the shaft. The first branch contained a little copper and was driven on a short distance. The second branch proved poorer. The lode itself was very poor where intersected. After driving a few fathoms eastwards another branch was picked up but although several fathoms were driven along it nothing of value was found. The Main Level was driven some distance south-eastwards. On the 23 fm. Level the crosscut to the west passed through the two branches mentioned above before reaching the main vein.

Apparently a good deal of driving was done on what afterwards were proved to be false veins, as Captain Francis was said to have eventually raised some £15,000 worth of ore from a parallel vein to the one first explored. In 1841 the mine was fully worked and employed up to 20 men but after the rich pocket was exhausted and all efforts to discover another had failed the mine was abandoned and has never been re-opened. When Captain Rowe made his report in 1878 on Roughton-gill he suggested that it might be found very profitable to re-open and develop at Carrock End but apparently the company was unable to finance such a venture and nothing came of it.

For working the pumping and winding machinery, water was brought by a leat from Carrock Beck. After driving the winding and crushing water wheel the water crossed the main road and was used to propel the pump wheel. Rods from this came under the road and across to the shaft where an angle beam changed the motion from horizontal to vertical, so lifting the pump rods in the shaft. West of the shaft two crosscuts have been driven to the lode but do not appear to have produced much ore. Further west two more have been driven but apparently also without success.

THE CALDBECK FELLS

Carrock Mine.

CARROCK mine, $4\frac{1}{2}$ miles south of Caldbeck, is one of the more recent workings in the district and is said to have been started by a Mr. Emmerson about 1854. All these workings were on Brandy Gill side of the valley and were started principally in the hope of finding workable lead ore.

The so-called lead vein at Carrock runs a little south of east dipping northwards along the flank of the fell through Poddy Gill and behind the old office and cuts through the Wolfram Veins. The occurrence of the large north-south quartz veins must have been known for generations but as they contained no sign of lead or copper the ancients had left them severely alone. Mr. Emmerson drove trial levels along the one which bears his name and also a trial in Brandy Gill which is on part of the Harding Vein. These operations came to an end in 1858 so presumably no worthwhile ore was found. It is said that the very large waterwheel which drove the Caldbeck Bobbin Mill was originally erected by Mr. Emmerson to crush and dress lead ore at Carrock.

Another party of adventurers discovered a lead vein higher up Brandy Gill in 1868. This vein runs east-west and extends about 200 yards between the Harding Wolfram Vein and one to the west which is probably the Smith Vein. Two levels were driven into the west side of the gill and at first some rich ore was found. It was only a small pocket and the venture came to an end. Some specimens of the rare stolzite were found in these workings. Years ago both wolfram and scheelite had been found in the North and South Veins but, apart from academic interest, were of little value to the then mine owners.

Trials were made by various people over the next 30 years but in 1901 operations for wolfram began in earnest and lasted until 1905. When restarted in 1908 by (it is said) a German company the mill was extended and the mine worked steadily until 1911. In 1913 the Carrock Mining Syndicate, under the management of the late Anthony Wilson of Thornthwaite, took over the property and there began a period of great activity. The mill was reconstructed and the latest type of concentrating tables were installed, increasing recovery to over 90%. The introduction of rock drills made it possible to raise the mill throughout to 50 tons per day and this production was maintained until the end of the 1914-18 war brought a slump in the price of tungsten; the mine closed in 1919. Eventually the mill was dismantled and the mine abandoned.

In 1942 the Ministry of Supply took up the property and began to re-open some of the old levels, also driving a new crosscut adit towards the Harding Vein. This work was started by a company of sappers from the Canadian Army and eventually there were also local men and a party of Spanish volunteers. When it became apparent that the submarine menace at sea was defeated it was decided that the erection of a dressing plant was not warranted and the mine was again closed down. Since then the price of tungsten has not reached a position where the Carrock Mine would be economic and so no further work has been carried out.

The geology of the area is complex and it is the only part of England, outside Cornwall, to yield tungsten in any quantity. The varieties of

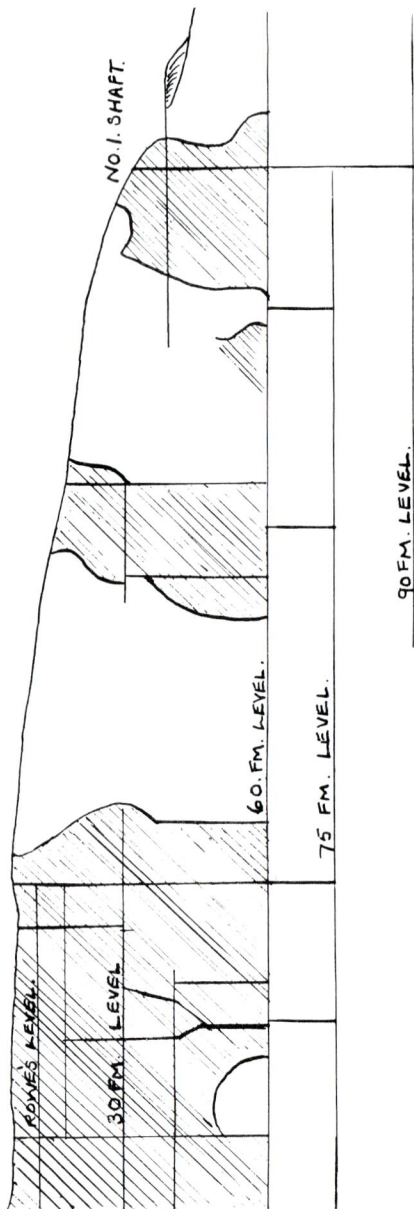

SW.

NE.

NO. 1. SHAFT.

ROWES LEVEL.

30 FM. LEVEL.

60 FM. LEVEL.

75 FM. LEVEL.

90 FM. LEVEL.

SCALE: 2.5 INS — 1 MILE.

Longitudinal section of part of Driggith Mine

THE CALDBECK FELLS

mineral specimens obtained are numerous, and more than 23 different species have been identified no doubt because of the complexity.

The Skiddaw granite outcrops near the junction of Grainsgill Beck and the Caldew River. Northwards the griesen comes in followed by schist and eventually the Carrock Fell gabro, which in places gives way to granophyre. South of the granite are Skiddaw Slates and northwards up Brandy Gill is a patch of decomposed granite. Still further north are the Borrowdale Volcanics.

The tungsten-bearing veins run north and south and generally hade to the west although the dip is very irregular and in places they are vertical. There are two major veins, Emmerson and Harding, but there are also many smaller ones, the chief of which are Smith's and the Molybdenite Vein, Smith's Vein is the most westerly of the group; it contains very little tungsten but much arsenic. Harding Vein has been the most prolific producer and therefore has the most workings upon it, having been wrought on both sides of the Grainsgill Valley. Emmerson Vein has also been worked quite extensively to the east side of Brandy Gill but trials to the south proved valueless.

The only working on the Molybdenite Vein is a small opencut, now almost filled with rubbish. No workings are now to be seen on the Lead Vein as these are buried beneath the dumps from the Emmerson workings. The Emmerson Vein proper is accompanied by a much later carbonate-type vein which is a considerable fault and to the north diverges westwards from the Emmerson. This carbonate vein continues through Dry Gill Lode and is seen in the first part of the Driggith 30 fm. Level, where it was found to contain a little copper ore.

Fine specimens of bismuth have been obtained from the Emmerson Vein and in places, accompanying this mineral, is a little gold. The tungsten occurred chiefly as wolfram but it is sporadically distributed through the very hard quartz and the output from this vein was inferior to the nearby Harding. A trial on the south side of Grainsgill did not yield any ore but some very fine quartz crystals were found in the loose and vuggy iode which was a complete contrast to the structure on the north side. In the main workings there was schist which caused difficulty in working and also in this ground the values fell away. Workings on the Harding Vein are extensive on both sides of the valley. On the Coomb side, five levels have been driven and a great deal of ground stoped out, but all these workings are now inaccessible.

The main workings on the northern side are from the Harding Level, which is on the same horizon as the main Emmerson Level and from which a horse tramway led around the valley to a level below the Penny Level on the Coomb side, all the ore then being tipped into a bin at the top of the Mill. There is also a short, lower Old Beck Level from the side of the stream. The New Adit Crosscut starts in the north bank, just below the mill site, and is 65 feet below the Harding Level; it cuts the vein at 250 feet in from the entrance. The crosscut continues to the Smith Vein at a further 150 feet. A little driving has been done on the vein and a rise put up to surface for ventilation. The tungsten values in the Smith Vein are very poor and the only mineral of any importance is arsenopyrite.

On the Harding, the North Drift has been driven some 540 feet and has proved the vein to carry workable values over most of this distance. At 190 feet north of the crosscut the lead vein passes through the quartz vein but shows only traces of galena. Much of the tungsten occurs as scheelite and is very rich in places, especially so in subsidiary veinlets on the west side of the Main Vein. There is also apatite which also glows in ultra violet light and is often mistaken for scheelite. Wolfram occurs throughout this part of the lode but is well-scattered through the quartz.

A ladderway rise at 330 feet north of the crosscut connects with the old Harding Level and provides ventilation. At the top of the rise the vein is in two parts with a large horse between, the rise top being on the western branch. A short distance northwards the veins join up again and continue strongly to the face which is still in ore. Near by some marmotite is seen and at first sight this can easily be mistaken for wolfram. North of the junction of the branches there has been no stoping but southwards to the level mouth the lode was stoped to surface in the 1914-18 war and is said to have been fairly rich.

South of the main crosscut the New Tunnel Level has been driven a total of 620 feet and proved ore-bearing almost throughout; it is not quite as rich as at the north side. Much of the driving is in the griesen, but the face has reached the granite. A rise connects with the Old Beck Level and some very rich wolfram was discovered at this point. No connections were made with the higher Penny Level, which extends a short distance further south.

How far southwards the tungsten values extend is not known, but float quartz containing wolfram has been found near Dry Swine Gill. Northwards, small trials have proved the vein up Brandy Gill and at the lead workings stolzite has been found, but it is not clear whether any wolfram or scheelite occurred. Quartz float on the Dry Gill side of the fell has yielded small specimens of bismuth but no tungsten. Although no specimen of tin ore has ever been found at Carrock the metal has been reported in some of the assays. So far there is no sign of deterioration in the veins in the shallow depth reached and very possibly they will yield similar ore at much greater depth. Their working depends on a higher price for tungsten than is usual except in time of war.

Driggith Mine.

DRIGGITH MINE is on the east side of High Pike, $3\frac{1}{4}$ miles south of Caldbeck, and has been opened on the South Roughtongill Lode, which here runs NE-SW and dips steeply to the north. The outcrop, worked extensively is at the head of Driggith beck coursing across the fell above the Red Covercloth at about 2,000'OD. Driggith is a comparatively recent discovery compared with Roughtongill and Red Gill Mines. The first mention was in 1790 when it was said to have been re-opened by a Mr. Rowe of Liverpool. Apparently the outcrop was quite rich, for it was opencasted for a considerable length and a shallow adit crosscut was also driven. The quantity of ore raised warranted the

Miners and timbering in the Harding Level, Carrock Wolfram Mine.

Driggith Mine, showing the 60 and 90 fm Levels.

Locomotive and train at the 90 fm Level.

THE CALDBECK FELLS

erection of a smelter by the side of Carrock Beck, quarter mile west of the *Horse and Farrier Inn,* and a stamping mill was built halfway down the valley and by the side of Carrock Beck.

How long it continued or how prosperous the undertaking was is not known as no records of output remain but it is said that the orey ground was all worked out to the 30 fm. Level by the end of the century. It is thought that some work was done below this level as Mr. Rowe is believed to have erected an engine— presumably a hydraulic engine — for drainage. By 1810 the mine was in the hands of Messrs. Richardson and Lowrey & Co., who erected a crushing and dressing plant below the 30 fm. Level in 1812. This plant had four sets of crushing rollersand was an improvement on the old stamp mill. In 1813 it was running night and day.

Work by this company continued until 1822. They are said to have obtained 200 tons of lead a year for at least five years and to have regularly employed 30 to 40 miners. Some of the lead is recorded as carrying 60 oz. of silver to the ton, but no doubt much of the lead was burnt off in smelting. Later figures show the silver content to be 25 to 30 oz. per ton of dressed galena running 75 to 78% lead. In 1822 R. G. Braddyll, of Conishead Priory, became the owner and started the Sandbeds or 60 fm. Level. He gave up in 1834 and is said to have lost a good deal of money.

John Barratt, of Coniston, followed and worked the mines for four years; then they passed to Messrs. Dickinson & Co., who had the Roughtongill Mine. This company worked until 1849 but are also said to have worked at a loss for several years. In 1849 Messrs. Brocklebank and Jefferey took over and Wm. Jefferey was appointed manager. He was a man of great experience and by re-organising the system of working underground and erecting a new, and at that time, modern dressing plant, was able to put the venture onto a solid financial basis. Zennor's rotatory buddles are said to have been first erected in this mill and were without a doubt a big step forward from the old-fashioned trunking boxes then in use. In the eight years up to 1857 some 1,950 tons of galena and copper concentrates were raised, the grade of the galena being 75 to 78% lead and 25 to 30 oz. silver per ton. The average labour force during this period was 35 men and 15 boys. In 1858 the 90 fm. Level was started from beside the dressing floors, being the lowest point at which the water from it would be usable in the plant.

Operations proceeded uninterrupted until 1870 when for various reasons it was no longer profitable to work and the mines closed down. Some work was restarted and continued to 1874 but probably only in the nature of a robbery; there are no records of output although some lead must have been raised. In later years the mine was split into two parts, the higher workings being still called Driggith, the lower part Sandbeds, and this has led to confusion in that the 60 and 90 fm. Levels at Driggith are the same levels at Sandbeds and not, as many people have believed, separate levels driven in from the Driggith beck.

In 1905, Mr. Goodall took up the Driggith part of the mines and after clearing the No. 12 and No. 30 Levels, erected a dressing plant near where Driggith and Dry Gill becks join to form Carrock beck near the site of the 1812 mill. This plant consisted of an ore bin with jaw-crusher, two sets of crushing rolls, trommel screens and elevator and two sets of jigs, all driven by a portable steam engine. A level tramway was constructed from the ore bin to the foot of the gill below the 30 fm. Level. An incline tramway was built up to this latter level so that the full cars could be lowered and at the same time the empty cars would be drawn up. At the mouth of the No. 12 Level a hopper and sorting grate was set up and the sorted ore was carted down to the mill.

Unfortunately the complex ore of galena, blende and barytes proved inseparable with the machinery then available, and operations ceased in 1907. Only 14 tons of zinc were sold and, although there is no official record of it, some lead and a little barytes were obtained. Nothing further was done until 1926 when the late J. H. Clay, who was working the Potts Gill Mine, attempted to clear both the 30 and 60 fm. Levels. It was found that most of the old timbers in the stopes had given way, letting down large falls of packed waste which were difficult and costly to clear, and eventually the venture was given up without ever getting to the central and richest workings.

In 1943 Messrs. T. & W. T. Shaw partly explored the 12 fm. and 30 fm. Levels but transferred the lease in 1944 to Mr. E. Gregory, who raised some barytes from the old outcrop workings. Messrs. McKechnie Brothers of Widnes, the recent lease-holders, bought out Mr. Gregory in 1948 and erected a dressing plant near the mouth of the 30 fm. Level. A little clearance work was done in the level and a rise driven through to the surface for ventilation, but the old stopes proved too difficult to re-work and the venture was not a success. All this plant has now been removed.

The South Roughtongill Lode is in two parts at Driggith which are there called the North and South Veins. In the western and central part of the mine they run together, but eastwards they split away and continue so up to the Great Crosscourse. Beyond this they do not appear to exist as so far no trace of them has been seen despite extensive trials and the large exploration made by the Sandbeds East Barytes Mine. These lodes are four to ten feet wide, mostly composed of quartz, but in the central part of the mine carry ribs of good barytes up to three feet in thickness. The chief lead ore is galena, generally confined to the middle part of the lode and it occurs both in ribs and as disseminated ore.

In common with other Lakeland deposits of lead, where there is a substantial quantity of barytes in the lode, the galena matrix is highly argentiferous. Near the surface there is a fair amount of pyromorphite with some cerrusite and specimens can still be found in the old open-cast workings. All the old stopes above the 30 fm. Level were re-worked in the 1860s for coloured lead ore thrown away by the ancients as valueless. Some chalcopyrite with a little chalcocite is also found in places in the lode and there has been some production of copper in the past. In the deeper workings a good deal of blende is associated with the quartz in which it occurs both as ribs and blotches and also as disseminated ore.

Several levels have been driven near the outcrop and of these No. 12 Crosscut has been examined in recent years.It is at the head of Driggith beck at about 1850'OD and cuts the lode at 70 yards in from the mouth and at 66 feet below the outcrop. A rise has been put up at this point to surface for ventilation and this is still open. The level has only been driven north-eastwards for a short distance with the lode in the face quite strong but showing very little mineral. Along the floor of the level and nearly up to the face there is a good showing of blende which is up to six inches of solid ore.

Just south-west of the crosscut a sump has been sunk 15 feet into a nice show of galena which still continues in the bottom. No attempt has been made to extract any ore below the floor of the level. It is thought that the 30 fm. Level would require to be driven some distance to come underneath it, although an old report of 1906 states that it was proposed to put up a rise to No. 12 from that level.

The level has been driven an unknown distance further south-west, as at 120 feet further it is blocked by a fall from an old stope. It possibly goes at least as far as No. 10 Crosscut, which is 800 feet south-west of No. 12. In No. 10 Crosscut the lode is five feet wide, consisting mostly of quartz with only traces of lead, and no driving along the lode has been done. This is the most south-westerly working.

The 30 fm. Level commences from the head of the right fork of Driggith beck and at approximately 1620'OD. It is extremely old, dating back to the 18th century. It starts directly on a north-south lode which hades west and is thought to be part of the Emmerson Vein of Carrock Wolfram Mine.It does not carry any wolfram minerals at Driggith but contained a little copper ore which has been stoped out above the level and also from shallow sumps below the level. This lode has been followed almost up to the Main Driggith Lode but it appears to get progressively weaker towards the north. The last part of the level leaves it and continues up to the lead vein as a crosscut, the total length being 660 feet. One or two small NE-SW strings carrying barytes are cut through after leaving the north-south lode. The crosscut next strikes a strong quartz lode some 3 feet thick with odd eyes of barytes but no lead is seen. This has been driven on south-west for 20 feet and north-east for 100 feet. A sump has also been sunk on the south-west side but is now filled with deads. It does however connect with lower workings as there is a flow of water down it in wet weather. The crosscut continues westwards and at 20 feet further from this vein cut the Main Driggith Lode. As far as is known workings extend south-west for up-wards of 1,000 feet and north-east for some 300 feet from the crosscut but in recent years only about 100 feet of them has been re-opened.

In 1906 the south-west level was cleaned to the face where it was intended to rise to No. 12 Level. The lode in the face is said to carry two feet of barytes and 18 inches of mixed galena and quartz. The north-east level was then open to the end where a rise was met with from the 60 fm. Level. In former times ore was tipped down and run out to the mill at Sandbeds. In 1906 the miners from Caldbeck used the ladderway by going in the 60 and out at the 30 instead of climbing up the fell.

The 60 fm. Level starts from Sandbeds side at about the 1,500 ft. contour and extends at least 3.000 feet from the portal, much of this distance is known to be stoped out above the level to the 30 fm. Level.

At a point 300 feet in from the mouth of the level a crosscut has been driven S.E. to explore for the Foxhole Vein but after passing through the Great Crosscourse the ground was found to be very disturbed and only very smashed and dislocated quartz was seen. Just north-west of the Great Crosscourse a crosscut has been driven due south for 250 feet in an endeavour to pick up the South Vein but no sign of it has been found. Several east-west stringers containing barytes, all of which dip southwards, have been passed through but none of them were strong enough to warrant further work. One stringer yielded some specimens of high grade copper ore but further work on it failed to find any more. The 60 fm. Level has been redriven up nearly to No. 3 Shaft and has proved all the old stoped ground to be in a state of total collapse.

At No. 2 Shaft there is a crosscut to the South Vein which at this point lies 60 feet south of the North Vein. To the south-west of the crosscut the level is blocked by fallen stopes but the short distance north-east was open to the face. The vein showed some barytes and the level has been driven a short distance further to see if it improved. Unfortunately it proved very poor and trials above the level have also proved valueless.

The central part of the mine contained the richest barytes ground and the 60 fm. Level has been re-opened primarily to see if any workable barytes does occur on the level in this area. The forehead of the new level is now about 300 feet short of the beginning of this area but in view of the bad state of the old stopes so far seen it is doubtful whether it is worthwhile to go any further and in all the ground explored no workable barytes has been found.

The 90 fm. Level was commenced in 1858 and proved a very bad investment for all concerned. Because of a surveying error, which was not discovered for several years, the level was turned too far east in an endeavour to pick up the North Vein and ran back into boulder clay, which at that point reaches 50 feet below surface. The level was then turned SW and followed a cross-course which was at first thought to be Driggith Vein. After driving some distance it was realised it was not the vein and the level was turned westwards to try to find the right vein. Eventually the Great Crosscourse was passed through and the ground changed immediately from soft, decomposed material to very hard greenstone.

A small lode was picked up but a re-survey in 1865 showed that it was far too far to the east to be Driggith Vein and the level was again turned westerly and passed through the Sandbeds Barytes Vein about 1,000 feet in from the portal. Eventually the right vein was located and the No. 1 Shaft was reached in October, 1867, after nine years work, during which 1,600 feet of level were driven. The level has now been re-opened to a face which is 600 feet SW of No. 1 Shaft and follows the North Vein only, no attempts having been made to crosscut to the South Vein. None of the ground explored has proved workable and the forehead looks particularly discouraging.

A 75 fm. Level has been driven SW from No. 1 Shaft for a distance of 1,200 feet up to No. 5 Shaft and therefore extended 600 feet further than the 90 fm. Level. Whether any ore of importance was found in this stretch is not known but it seems unlikely as none has been worked.

In the long time taken to reach the No. 1 Shaft all the worthwhile ore above the 60 fm. Level had been exhausted and in the sinkings put down below, the level water proved troublesome and costly. Actually No. 1 Shaft was sunk by hand to the 90 fm. Level before that level reached the shaft. The 75 fm. Level was also driven up to No. 3 Shaft on the hanging side of the vein following the small rib of barytes, all the material from this work being wound by hand to the 60 fm. Level. When the connection was made to No. 1 Shaft it was confidently expected that stopes above the 75 fm. Level would yield a great deal of lead but on breaking into the lode proper it was found that it now contained a great deal of blende and that what galena there was proved to be very disseminated and much more difficult to dress. The long struggle with the 90 fm. Level and the large sum eaten up by the work done below the 60 fm Level finally caused the company to go bankrupt. The mine was closed for good in 1874 after nearly a century of continuous operation.

There now seems little doubt that the rich ore found in the upper part of the veins is in fact a zone of enrichment caused by the decomposition of the ground and when the hard, unaltered greenstone is reached there is a gradual falling off in values until in a short depth the veins are practically barren.

Why the orebodies on Caldbeck Fells, particularily in the Borrowdale Volcanics, should differ so greatly in character to those at Patterdale and Coniston is difficult to understand and apparently baffled the old men also; otherwise they would not have attempted to explore at depth.

Braefell Mine.

THE Braefell mine lies on the west side of Roughtongill Beck about a quarter mile south of the site of the smelt mill. The hillside is fairly steep and thickly covered with boulder clay through which small becks have carved deep ravines. The northerly and larger one of these is called Ramp Gill but it does not appear to have reached solid rock at any time. The southerly cutting, now devoid of water, is called Dry Smale Gill and between them the ancients have made a "hush" which cut through to bedrock and so exposed the lode. This runs diagonally across the flank of the fell on a bearing of WNW and hades to the south. It appears to be mostly composed of quartz with some galena, pyromorphite and traces of malachite.

Two levels have been driven along the lode at 10 fathoms apart with the upper level also approximately 10 fathoms below the outcrop in the hush. At this point a rise has been put through to surface, no doubt for ventilation and possibly also on a bunch of ore. From the size of the dumps there is no doubt that both of these levels extend a considerable distance but they have long been closed up and it has not been possible to explore them. No plans are in existence and there is no idea of the extent of the stopes. It is not known when the workings were made but in 1866 the Red Gill Mining Company had two men cleaning out the levels for examination. Nothing further was done and presumably no ore of value was seen.

Below the lower level is the site of the old dressing mill which seems to have been worked entirely by hand as there is no trace of any water-wheel pit. Very probably the mixed ore would be carted up to Roughten-gill to be crushed and washed; apparently the ore from the mine was tipped down a wooden chute and sorted from a little bin at the bottom. The pieces of ore were broken to hand-jig size on a large iron plate which is still there. This process was known as "bucking" and special hammers which had a curved face were used. The coarser ore was then hand-jigged and the fines cleaned in box buddles, or "trunking boxes", as they were called in Lakeland.

How much ore was actually got is unknown but the work was probably done when Roughtongill was prosperous.

Smale Gill Veins.

IN Dry Smale Gill a strong quartz vein carrying lead and copper was discovered by the ancients. This vein runs nearly east-west and for much of its course follows the gill but at no place can it actually be seen in situ and its hade is not known. At the foot of the gill, just below the small sheepfold, a level was made in a westerly direction and was presumably the main working. A strong stream issues from the level, the mouth of which has long ago collapsed. The old men had a hand-washing and sorting floor and the remains of the ore bin can still be seen. Some fair pieces of mixed lead-copper ore can be picked up and specimens show small quantities of linarite and chrysocolla.

Just below the Red Gill road is a run-in rise which comes up from this bottom level. Thirty yards above the road is an old level running 10°W of N but the entrance fell in long ago and only quartz is seen on the tip. Still higher up the gill are two more levels but no ore is to be seen in either tip. The highest tip contains some ankerite with the quartz. It appears that this vein carries some ore deeper down but there are no records of output and it is doubtful if any of it was considered payable.

The vein in Wet Smale Gill is also a strong quartz lode which runs in a north-westerly direction. It has been tried about halfway up the gill from a small sinking but no ore is to be seen in the quartz. Several more trial pits have been made across the flank of Brae Fell, apparently without success. Just above where the old water race to Red Gill Mine leaves Wet Smale Gill a very old level has been driven southwards into the bank. Small pieces of barytes are to be found in the tip and there are also small amounts of malachite but no sign of the vein is to be seen.

Hay Gill Mine.

THE old copper mine of Hay Gill is thought to date back to the Elizabethans and is reputed to have yielded some rich ore, but this was long before any records were kept. It is difficult to visualise the extent of the workings as no plans or section appear to have survived. The Main Lode runs N50°E and crosses the stream about 100 yards below the junction of Long and Short Grain Becks. The country rock is the Eycott Lava which is one of the lowest members of the Borrowdale Volcanic series on Caldbeck Fells.

An old working, probably the original and long ago collapsed, runs into the southern bank of the gill. On the northern side there is an old level, also long fallen in and, by the side of the stream, the collar of an old shaft. This shaft is sunk in the footwall side and has been drained by a 20ft. diameter by 3ft. breast waterwheel, of which only the stone-work remains.

Some 200 yards lower down the gill is the entrance to the No. 1 Crosscut which runs south-east for over 500 feet to cut the lode and is then driven north-east along it to the shaft which it drains at 60 feet below the collar. This level has been the principal working of the old men. Remains of a sorting floor and odd pieces of good grade copper ore can be found. Much of the waste from the workings must have been washed away down the gill as the present heaps bear no relationship to the extent of the workings. How far the shaft goes down below this adit is not known, nor can we visualise the extent of the areas stoped. In 1866, when the property was in the hands of the Mid-Cumberland Mining Company, Mr. Vercoe employed three men to re-open this level and examine the old workings but whether they succeeded is not known although it is possible because in 1871 some 10 tons 4 cwt of ore, running 15¾% copper, were sold.

Lower down the gill, at a point 1,100 feet from the level mouth and 120 feet below it, a Dale Beck Level was started in 1861. This level was driven SSE and followed a small barren vein for a distance of 81 yards, but was still a long way short of striking Hay Gill Copper Vein.

In 1859 there was an output of 23 tons 1 cwt. of dressed copper ore from Hay Gill but from where it was obtained is obscure because the main working collapsed in 1866. The recorded output in 1860 was only 1 ton 11½ cwt. of dressed ore. By 1874 all operations had come to an end and the rails and other plant were up for sale. Since then nothing further has been done.

The China Clay Mine.

SITUATED on the north side of Great Lingy Hill near Hare Stones, and about 1½ miles south-west of Potts Gill Barytes Mine, is the Caldbeck China Clay Mine. The deposit is connected with the Roughton-gill South Lode which here is almost devoid of lead and zinc minerals but contains umber in rotten sugary quartz three to six feet wide. This lode runs NE-SW dipping NW but is not very well defined in the workings. The hanging wall is Borrowdale Volcanics but the footwall is decom-

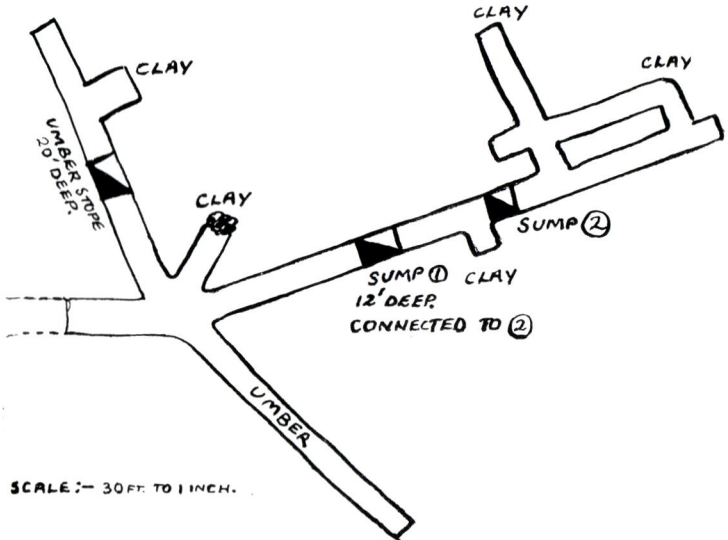

A plan of the China Clay Mine.

posed hypersthenite and is a type of china clay. A sample obtained from the workings in 1942 had the following analysis:

SIO_2	75.60	
FE_2O_3	0.20	
AL_2O_3	18.30	
CAO	nil	Moisture in
MGO	0.25	the sample 15%
H_2O+	5.65	
Loss on ignition	nil	
	100%	

The mine is entered by a short crosscut adit at approximately 1800'OD. This adit runs due south for 60 feet, at which point the South Roughton-gill Lode is cut. Two drifts have been driven along this lode, one to the north-east for 51 feet and the other south-west for 52 feet. There is a small under-hand stope 20 feet deep in the NE Drift from which umber has been obtained. There is very little umber in the SW Drift and no signs of lead or zinc in either place.

The crosscut advances 106 feet south of the lode, all in china clay which contains small pockets of umber. The end of the drift is filled up but probably does not extend much further and very likely has reached the limit of the clay.

As can be seen from the plan, side drifts have been started and the clay is especially good towards the north-east. These workings were made in 1887-1894 in connection with other umber and barytes works at Roughtongill Mine and were connected to the latter by an aerial ropeway. No figures of output are available but obviously several hundred tons of the clay have been raised.

Potts Gill Copper Mine

AN old copper mine just south of Potts Gill Farm has worked an east-west copper bearing lode; two adits have been driven. The upper or No. 1 Level starts directly west along the lode from the side of the gill at exactly 1,000'OD. The lower or No. 2 Adit is a crosscut 60 feet below No. 1 and has been driven SSW for about 300 feet to reach the lode. This level is walled up and used as a water supply for the farm. An attempt was made about 20 years ago to re-enter the No. 1 Level but it was found to be totally collapsed and the work was eventually given up before the actual workings on the lode were reached.

Below the mouth of this level are the remains of the old sorting floors and sheds and some pieces of rich chalcopyrite may still be found lying about. Eastwards, the lode has been tried from open cuts across Nether Row Brow. At the sharp corner in the road to the barytes mines another cross-cut has been put in, but apparently without success. No records or plans of the old workings are available and the actual extent of them is unknown. It is not thought likely that much workable ore occurred. There is some barytes in the lode but not enough to be worth working on that account.

The Upper Level is very ancient and was re-opened in 1868 by the Mid-Cumberland Mining Company. The Lower Level was commenced the same year and operations were vigorously pursued, up to 20 men being employed, but by 1871 the mine was abandoned, no worthwhile ore having been found.

Caldbeck Mines.

THE Caldbeck group of mines, worked exclusively for barytes, stretch for two miles across the northern slopes of High Pike. From east to west they are: Sandbeds East, Sandbeds West, East Potts gill, Old Potts Gill and Deer Hill.

Barytes mining started at Old Potts Gill in 1844, but very little was done and there are no records of output obtained. After some years it was restarted and was working in 1880-81, but it closed soon afterwards. The mine was re-opened by the Cleator Moor Iron Company, who had Roughtongill and the China Clay Mine, from 1887-92. Very little further work was done until the 1914-18 war, when the mines were re-opened by Mr. Goodall, being taken over by the late F. G. Ryland in 1915 under the name of Potts Gill Barytes Mining Company.

During this period there was a dressing and grinding mill at Caldbeck and work proceeded steadily until 1920, when the slump in the price of barytes (due to cheap importation from Germany) brought the operations to a close. The mines passed into the hands of the Caldbeck Mining Company, who built a small dressing plant below the No. 1 Adit Level and sold the barytes in a washed state, most of it going to Barnsley for grinding.

Eventually Caldbeck Mines Limited was formed and a deeper level known as the Endeavour Level, was started in 1928. This level is at 1319'OD and is a crosscut, 1.500 feet in length, to the main lode which it cuts at 165 feet below the upper adit. The crosscut was not completed until 1938 under the management of the late E. T. Borlase. Once the lode was cut and opended up sufficiently to start production a dressing plant was built. In 1940 the mines passed to Messrs. McKechnie Brothers Limited of Widnes and since that time they have been thoroughly explored and worked over the full length of the lodes.

The total known production from the group since earliest operations amounts to 140,000 tons (August, 1965) of dressed barytes. The total aggregate length of main levels is nearly eight miles, of which main crosscuts total over $1\frac{1}{2}$ miles.

The barytes-bearing lodes cross the northern flank of High Pike in a WNW direction and hade southerly at generally 65° but are often much flatter and in places vertical. The workable barytes-bearing lodes are confined to a narrow belt of country which has been tried for two miles in length. Long crosscuts to the north and south have failed to pick up any further workable lodes although other weakly-mineralised ones have been found. Of these, the Rainbow Lode has been tried for a length of 1,150 feet along the strike. It often looked very promising, but no ore of importance was discovered. This lode has, however, an east-west direction and carries traces of galena over all the length explored. There is in parts a good deal of manganese soot or wad; a great deal of quartz exists throughout.

Ropeway and sand ponds at Pottsgill Mine.

THE CALDBECK FELLS

The lodes lie entirely in the Higher Ireby Group of the volcanic series of Borrowdale which dips steeply south and are most productive where the country rock is oxidised. The lodes are generally pinched and barren where the country is unaltered. They vary greatly in width from a mere string to 20 feet; the average is probably about three feet. The barytes content also varies greatly, and it is calculated that he average width of the barytes in the ground worked has been 1′ 8″.

Only small amounts of other minerals exist in the lodes apart from quartz, and this is mostly a loose and cindery variety which is very common. Other minerals found are pyrolusite, psilomelane and wad, ankerite—both dark and light coloured varieties, odd specimens of galena, copper in the form of chalcopyrite, chrysocolla, malachite, chessylite, cornwallite, cuprite and the black oxide—melaconite. The assays have at times shown up to $\frac{1}{2}$% $BaCO_3$ but no specimen of witherite has been found. Examination of the lode with the ultra violet lamp has shown spots of witherite in some parts of the solid barytes lode but it could not be discerned with the naked eye in ordinary light. A form of stalagmite barytes was found in the 90 fm. Level workings of Sandbeds West Mine which contained 57.53% $BaSO_4$ and 33.80% SiO_2.

Old Potts Gill Mine.

OLD POTTS GILL, the original barytes mine, was started possibly about 1844, with the first workings westwards from the side of the gill and directly on the Main or North Lode. It was known as No. 5 or Potts Level and was worked through to surface for the first 200 feet of length with the total length of the level being about 480 feet. When the mine was re-opened in 1891 by the Cleator Moor Iron Company the No. 3 or Nicholson's Crosscut was driven to get a further 22 feet of height below No. 5 Level. No. 1 Crosscut, which is 26 feet lower than No. 3 was driven at 1484′OD for a distance of 240 feet to cut the lode in 1914 by Mr. Goodall.

The South Lode had also been discovered and a shallow level called Hodgson's Level had been started westwards from the side of the gill. The No. 1 Crosscut was extended a further 140 feet to cut this lode, and a rich little stope was worked through to the surface. Further east on the fellside a short level named Angus Gulf was driven on the South Lode and a good pocket of barytes extracted. When Mr. Clay formed the Caldbeck Mining Company and took over the mines he extended the No. 1 Crosscut a further 160 feet south but nothing of importance was found. The No. 1 Level was driven west a total of 710 feet from the crosscut and opened into a very fine seam of ore about 300 feet in length, and this was stoped to surface. Eastwards, the level was driven 220 feet from the crosscut and reached the Gill Crosscourse. On the South Lode, levels were driven about 130 feet each side of the crosscut and the ore stoped out.

In 1928, when Caldbeck Mines Limited was formed, it was decided to commence a long adit crosscut from the west side of the gill at 1319′OD or 165 feet below No. 1 Level. Named the Endeavour after Lipton's America Cup challenger, it was aimed at the centre of the ground worked in No. 1 Level and was driven in some 850 feet before lack of funds caused it to be abandoned. Work restarted in 1936 under the management of the late E. T. Borlase, and the face was turned more westerly, into softer ground. After driving a total of about 650 feet further the lode was cut. Some driving was done each way and a good deposit of barytes was laid open. A rise was driven all up to the No. 1 Level for ventilation and this showed good ore all the way through. The Endeavour Crosscut was extended a further 30 feet and struck the South Lode, proving that it joined the Main Lode further west. A shoot of ore averaging 30 inches wide and 200 feet long was opened out on this lode.

A trial on surface, about 400 feet west of the main ventilation rise from the west side of No. 1 Level, struck a very rich deposit of barytes which was, however, accompanied by much quartz and manganese, both as soot and hard, solid ore. Eventually No. 1W Level reached a point 1,330 feet west of the No. 1 Crosscut, with the face showing up to one foot of barytes. There was a great deal of soot in damp, soft, rotten rock. Trouble with the manganese in the mill brought this part of the mine to a halt in 1942.

The system of working the ground in this mine was by "cut and fill" stoping, which was more costly than the "sub-level" system later developed and used entirely in the other mines. The ore had to be richer to carry the cost of filling the topes with waste and building up the timber ore passes, or "mills", as they were called. Trials of the sub-level system were made in suitable ground and it was so successful that it was adopted. In all fairness, howerer it must be stated that the ground worked in the mainstope from the Endeavour was very loose; the ore could only be recovered by the cut and fill system. Very little of these conditions have been seen in the rest of the mines and so it was possible to stick to the sub-level method.

By 1947 work had ceased in the Endeavour Level as all profitable ground had been exhausted. Nothing further was done in this area until 1961, when a trial adit was driven on Deer Hill about 500 feet west of the No. 1W workings. This crosscut struck a strong lode which showed up to three feet of barytes were passed through. Some digging further west also revealed traces of a worth-while vein and it was decided to drive a 500 feet crosscut to the lode at a point 520 feet west of the trial adit and 120 feet lower down the hillside. This crosscut is at 1414'OD and is driven 10°W of S for 515 feet to the lode which shows three feet of barytes were intersected. The crosscut continued 65 feet further south; after passing through a small copper-bearing vein at 55 feet, the face of the crosscut is in hard unaltered rock.

The level was driven eastwards 340 feet where this rock again appears and the lode becomes hopelessly barren. A rise was put up to surface and a level off it at the horizon of No. 1 Level was driven east for 320 feet. There was very little barytes in this level but at 100, feet back from the face another rise was put up to the trial crosscut. This is 60 feet above and there was ore all the way. A great deal of soot and loose quartz lay in the lode which is, on the whole, very wide and poor. All the workable ground has been removed and the workings are now abandoned.

In the whole of the Caldbeck Mines workings the Potts Gill Mine has produced the greatest variety of minerals as well as the best crystaline specimens of barytes. On the South Lode, just west of the No. 1 Level Crosscut, a small north and south vein came in on the hanging wall side and a little bunch of copper ore was taken out. The bulk of the ore was chalcopyrite and the black oxide, melaconite, some malachite and the rare cornwallite. In the Endeavour Level a small "lead" vein was passed through; it contained arsenarpyrite and marmatite besides a very little galena. Manganese was plentiful in the western part of the mine, both as wad or soot and psilomelane.

On the hanging wall side of the main ore pocket in the Deer Hill Level some specimens of cuprite were obtained, but this appears to be very rare. None has been seen in any of the other workings. Odd stones with small amounts of crysoscolla have been seen in various parts of the upper workings. On the East 80 ft. Level of the East Pottsgill workings, specimens of mineral leather were obtained from a patch of hard, un-altered rock on the hanging wall side of the New North Lode.

East Pottsgill Mine.

IT was decided in 1942 to make some trials on the East fell at Potts Gill and two levels were started, one on the South Lode and the other on the Main or North Lode. Both levels were at about 1600'OD and began directly on the veins.

The Thomas Level on the South Lode was commenced from the end of an old working known as Angus Gulf and, after driving forward 50 feet, a large cross-course now called Shaw's Crosscourse, was struck. This runs NNE and dips to the east; it appears to be a large downthrow, although the lateral displacement of the lode was no more than six feet to the north or left hand. Driving on the lode was continued for some 450 feet and very little workable barytes was found, but the rock became progressively unaltered as the working advanced eastwards.

The Jon Level on the Main Lode lies 220 feet north of Thomas Level and is at 1594'OD, starting on a quartz lode two to three feet wide with very little barytes up to Shaw's Crosscourse, which is cut at 130 feet in from the mouth.

A crosscut north, just east of the crosscourse, pierced a fine lode called the New North Lode at only 20 feet and this proved to be a much better proposition than the Main Lode because there was a continuous run of excellent ore for 750 feet. By then the workings were passing through the crest of the hill and beginning to lose height. The lode began to split up into stringers, none of which had much strength. It is doubtful if it persists much further as subsequently it was found that there is a great barrier of unaltered rock ahead which comes out to surface. At a point 780 feet east of the crosscourse there is a junction with a NE-SW vein which has been called Blockley Lode after its discoverer. This was driven on for 500 feet and after being stopable for the first 360 feet ran into hard, unaltered rock and was completely barren.

All the ore on both these lodes was worked right through to surface, a height varying from 90 to 150 feet. Two other tiny crosscourses were passed through in the drive on the New North Lode, one at 355 feet east of Shaw's Crosscourse and the other at 370 feet further east. Neither moved the lode more than its own width and both throws were to the north. A further level called the Victory was started from a poin 310 feet west of Shaw's Crosscourse and 30 feet below the Jon Level. A rich bunch of ore was discovered immediately east of Edgar's Crosscourse, with barytes and quartz up to 18 feet wide and much umber or soot, the crosscourse itself being eight feet wide and also containing a large amount of soot and rotten rock. Unfortunately this deposit proved to be only a pocket and very little more ore was seen until Shaw's Crosscourse was passed. At a point 290 feet east of the level mouth a short crosscut north penetrated another unknown lode now called the Back Vein and this was driven on eastwards for 180 feet to where it joined back into the New North Lode. The drift on the latter vein was extended to 670 feet from the mouth of the level and at 470 feet in a crosscut south was driven 150 feet to the Main or North Lode but this proved too poor to work.

The Gill Level at 1520'OD was started from the side of Potts Gill opposite the old No. 5 workings and driven in the footwall side of the vein until the Gill Crosscourse was passed through at 140 feet from the entrance. The level then followed the lode which was barren up to Edgar's Crosscourse which was met with in a further 130 feet of driving.

The rich bunch of ore found at the entrance of the Victory Level proved to be very much restricted on the Gill Level, the length of payable ore being only 45 feet and very narrow on the level itself.

At 510 feet from the entrance, Shaw's Crosscourse was passed through and, as above, the level was turned onto the New North Lode. From this a crosscut was put out north to the Back Vein, which was found to be very rich, completely filled with barytes up to six feet in width. Quartz was absent. It was driven on for 270 feet, where it re-joined the New North Lode.

At 30 feet east of the North Crosscut a crosscut south was driven for 185 feet and cut the North or Main Lode at 80 feet. This proved workable and was driven on back west towards Shaw's Crosscourse for 100 feet and stoped out for 35 feet above the level. Eastwards the lode was followed for 500 feet, where it cut the Blockley Lode, but the rest of the ground proved unpayable apart from lengths of 80 and 30 feet. At 150 feet east of the crosscut a rise was holed through to the South Crosscut on the Victory Level and stoping was continued for a further 75 feet above this level.

The drift on the New North Lode was in almost continuous ore for 750 feet past the South Crosscut to the Main Lode. The Blockley Lode was cut at 690 feet and a 140 ft. crosscut was driven to the latter vein from a point 490 feet east of the South Crosscut. The drift along the Blockley Vein joined up with the level on the North Lode and was continued for 105 feet along the Blockley after the intersection but it proved to be quite barren in hard, unaltered rock. There was no sign of the Main Lode to the east of the Blockley Vein. Stopes on the Blockley were worked up through to the Jon Level over a length of 320 feet but the lode below the Gill Level was poor.

In 1946 it was decided to drive a long crosscut to get below these workings from near the mill and at the same horizon as the Endeavour Level, 1319'OD. This crosscut tunnel was driven south-east and cut the New North Lode at 1,940 feet from the portal. Much of the tunnel is in hard, unaltered rock and no other lodes of workable size were found, though it passed through several strong veins and at least two of the crosscourses were seen. The Rainbow Lode was cut near the entrance but this also proved of little value. In driving west along the New North Lode there was an unfortunate accident to two brothers, one being killed and the other badly hurt. The accident followed a fall of roof out of a large loose vugh, which caused the timber to collapse.

At a point 350 feet west of the tunnel a crosscut was driven south for 105 feet to reach the Main or North Lode and this was followed west-wards for a further 310 feet. A stope 110 feet long was worked up to an intermediate level (Turner's) 100 feet above. The only ore on the New North Lode at Tunnel Level was in two stopes at the western end one 40 and the other 140 feet long, which reached up to Turner's Level out of 875 feet of ground explored, most of it being in hard, unaltered rock in which the lode was completely barren. On the Intermediate Level, stopes on the New North Lode totalled 360 feet in length, about half the length of the ore on the Gill Level.

On the Main Lode the length of ore was 610 feet on the Intermediate Level, which was a large increase on the Gill Level, but most of this was very poor in the floor for, as stated, only a length of 110 feet was worked on Tunnel Level. The Main Tunnel was continued south-east for 160 feet, being turned round towards the east to cut the Blockley Vein but this was not found. There is a flow of approximately 150 gallons of water per minute from these workings, even in the driest weather.

A crosscut south was driven on Thomas Level; it started a few feet east of Shaw's Crosscourse. This crosscut was 1,030 feet in length and passed through three veins, only one of which contained any barytes worth following. This was the Job's Well or Edgar's Vein at 630 feet along the crosscut. It was driven on for 130 feet west of the crosscut towards the crosscourse, a little ore being raised from trial rises. The next vein met with at 200 feet further south was opened out for a total of 25 feet but contained mostly quartz with very little barytes. At a further 90 feet a strong vein, mostly of quartz with a little barytes was cut. This ran NE-SW dipping NW, and was thought to be the continu-ation of the Blockley Vein. It was driven on for a total of 75 feet but showed no signs of improvement. The end of the crosscut is in rock bordering on the unaltered type but if it had been continued in a SE direction for approximately 550 feet further it would have cut the large North Roughtongill Lode which has been tried on surface immediately above this point. The trial shows the lode to be mostly of hard quartz with a little galena and splashes of barytes.

In 1938 the dressing plant was constructed using second-hand plant, mostly from Cockhill Mine near Pateley Bridge. The plant was set out on old-fashioned lines, all the machines being belt-driven by flat belts from main and secondary shafting, and a 35 H.P. Blackstone horizontal diesel engine was the prime mover. McKechnie Brothers Limited took over in July 1940 and a new compressed air plant, consisting of a 75 H.P. Crossley horizontal diesel engine belt driving a 250 cu.ft. per minute at 100 lbs. per sq. inch Alley McLellan "Sentinel" compressor, was installed, making it possible to have four rock drills in regular operation. In the next few years a great deal of ground was explored and opened up.

At the same time several additions and alterations were made to the mill to cut down labour and increase the output. Other alterations further improved the process but basically the system was the same as it would not have been possible to achieve anything better without building a complete new mill. It was, therefore, a Heath Robinson affair but the results from it have been remarkable considering that most of the plant was aged and the major alterations were accomplished during the war without stopping production.

The plant had a capacity of 200 tons of crude ore per five day week, with a running time of about 40 hours. Depending on the grade of ore treated, the output of dressed barytes has been generally 100 tons per week but it varied to as high as 150 tons and as low as 70 tons with the same throughput. Every effort was made to control the grade of crude ore so that it averaged at least 50% barytes. This was found to be the lowest grade at which a 92% $BaSO_4$ concentrate could be made with maximum throughput.

An aerial ropeway of Monocalbe type, 1,300 yards long, took the dressed barytes down to a 200 ton bin at Nether Row. There was a standby compressor set consisting of an 80 H.P. horizontal Crossley diesel engine belt driving a 200 cu.ft. per minute Ingersol compressor and the mill was latterly propelled by a 55 H.P. horizontal Crossley diesel engine. The smithy was equipped with a grinder, drilling machine and drill sharpener, but this last item was only used for re-shanking steels as latterly all drilling was done using Tungsten Carbide steels and also Tungsten Carbide detachable bits. The office was a 30 x 18 feet wooden building divided into two offices, a store and a change room.

The total output of the Potts Gill Mines (Old and East) has been at least 90,000 tons of dressed barytes, which equals a solid rib of barytes 1′ 4″ thick over the whole area stoped. There is some ore in the floor of the Endeavour but with a flow of water of 100 gallons per minute it would not be possible to take it out economically. All the workings at Potts Gill are now abandoned and are not likely to be re-opened.

Sandbeds East Mine

THE outcrop of the barytes lode at Sandbeds was found in 1927 by John Hemmingway and William Blockley, of Mungrisdale. They were prospecting for Mr. J. H. Clay. Nothing was done to open it up until 1946 when a crosscut 870 feet long was driven south to it from the side of the mines road at the horizon of the old Driggith 90 fm. Level. This crosscut struck the lode at a point 160 feet below the old prospecting trial trench and, where cut, the vein showed about two feet of

Barytes hoppers in Sandbeds Mine.

barytes. A rise was put up to surface for ventilation and a second way out. Driving was then resumed east and west along the lode.

The eastern drive opened into a rich vein of ore which was more or less continuous over the whole 1,060 feet driven, at which point the level came out to surface in Blea Gill (not to be confused with the Blea Gill at Roughten Gill Mine). Westwards, the lode did not seem to have the same characteristics as that seen in the eastern level; it was so unlike it that there was doubt that it was the same lode. When crosscuts failed to discover anything else running parallel it was concluded that it must be the same. Driving was continued along the lode for 735 feet, at which point the large Sandbeds Great Crosscourse was struck.

The 100 fm. Level was started at the foot of the hillside to the west of Blea Gill with the first 40 feet as a crosscut south, at which point the lode was cut. The level was then driven westwards for 1,780 feet, of which 870 feet was rich enough to stope out to the 90 fm. Level. Generally the lode proved very poor indeed and a great deal of the level was driven through unaltered rock which showed that the working was at the base of the barytes zone. The most easterly working is the How-thwaite Level which lies by the south side of the mines road, about $\frac{1}{4}$ mile from Calebrack Farm and at 1125'OD. This level starts almost directly on the lode and has been driven westwards along it for 480 feet. The lode is generally large and in places contains a good deal of quartz, but nowhere along the length explored is there any workable barytes.

A feature of the East Sandbeds Mine was the great thickness of loose glacial till which, in some points, reached 50 feet below surface and caused much trouble in the stopes. In some places a good deal of the ore from pillars was lost because this loose material ran into the stopes. The total area stoped amounted to 4.45 acres, which yielded 16,964 tons of dressed barytes, equal to a solid rib of barytes 10 inches thick. Thus on the whole the yield was very poor. All the workings are now abandoned and closed up.

Sandbeds West Mine.

SANDBEDS West is a continuation of the East Sandbeds Lode and it was not discovered until 1956. A trial crosscut had been made into the fell towards Pottsgill at about the same horizon as the Driggith 60 fm. Level in 1952 but had been given up when no lode of importance had been found. In July, 1956 a survey of the Intermediate Level above the East Sandbeds 90 fm. Level West was made and the projected position of the lode in relation to the 60 fm. Crosscut was worked out. This showed the lode to lie south of the face of the crosscut, which was then restarted. It cut a very fine show of barytes on September 14th.

Eventually this 60 fm. Level reached a point 1,130 feet west of the crosscut but only 570 feet of this distance proved stopeable. At about 300 feet west of the crosscut the lode became filled with dolomite and quartz, with only small pockets of barytes. These were of high grade water-clear spar. There was also a good deal of manganese soot and some of the dolomite was of the very dark variety which carries approximately 12% of manganese. The country rock in this area was unaltered and these conditions existed for a further 300 feet. The lode then turned north about 10° and the rock gradually became softer until at 900 feet west of the crosscut some very good barytes was encountered, much of it as detached masses of several cwts. embedded in manganese soot. Unfortunately this soot was only about 100 feet in length, after which the unaltered rock again appeared and the lode was entirely barren, being only a few inches wide, filled with crushed rock and very wet indeed.

At a point 120 feet in from the mouth of the 60 fm. Level a crosscut north was driven 12° W of north for 550 feet, at which point it struck washed out ground in a gill. Rather than drive through this gravel in case the drift passed over a concealed lode, a branch was turned off to the NW about 30 feet back from the face and continued for 200 feet where a rise was put up to surface for ventilation. The crosscut was then turned back on to line and at 240 feet further on it pierced a small lode, beyond which it was continued for 170 feet without finding anything more. The rock was also changing into hard, unaltered greenstone and looked very unfavourable.

The small lode, known as the Dumpy Vein, was tried for 115 feet to the west and 20 feet to the east of the crosscut; it was very narrow though persistent and contained only a little barytes with a great deal of manganese soot. A two foot quartz lode running NE-SW was found near the foot of the ventilation rise but it was quite devoid of any minerals.

The 50 fm. Level was driven west a total of 1,370 feet and of that length 760 feet proved stopable. The level encountered the unaltered rock at 400 feet in from the mouth, but at 600 feet the brown rock came in, only to go out again at 830 feet. It persisted until about the 1,000 feet mark. Beyond this point the lode opened out into a very rich bunch which extended 200 feet before being cut off by the unaltered rock. This ore cut out almost at the horizon of the level but widened rapidly upwards, reaching as much as eight feet of solid ore in the central part. The lode was very flat in this ore body but it had a hanging wall of unaltered rock; this extended nearly to the surface and made it possible to extract every particle of this splendid bunch.

The 40 fm. Level was driven a total of 1,350 feet along the lode and proved richer than the preceding level in that 850 feet of length proved worth stoping. The 30 fm. Level extends 1,100 feet along the lode and of this distance 650 feet has proved payable and stoped out to surface. At a point 550 feet in from the mouth of the level a branch is thrown off to the north and followed for 330 feet to where it is down to a mere string and turning back south rapidly to join the Main Vein, which it does at about 100 feet further west. The branch was tried on the lower levels but pinched out in depth, and only on the 40 fm. was any ore found. At the face of the 30 fm. Level a NE-SW quartz lode was struck. The country rock was very hard and unaltered, and so very little trial was made on it. No attempt was made to pick up the Main Lode through the quartz lode.

The 20 fm. Level was driven 525 feet to reach the same quartz lode, which again proved quite barren although specimens of arsenopyrite with specks of marmatite were obtained from the western side of it. The barytes lode was picked up to the west and driven on for 65 feet. It was very narrow, the rock hardening and becoming so unpromising that work stopped. A little ore was stoped above this part but it did not extend upwards to the surface. This face was the most westerly point reached and left a gap of just over 400 feet to the nearest Potts Gill workings.

The 67 fm. Level Crosscut is 420 feet long and has been driven from the lowest point at which water could still flow to the tank supplying Hesket Newmarket village. The lode at the point where it is cut was in two parts, with one foot of barytes on the footwall followed by a horse 15 feet and then up to three feet of barytes on the hanging wall. East of the crosscut the hanging wall was followed for 135 feet where a small crosscourse was encountered. The level was turned SE and broke into a stope from the Intermediate Level of the East Mine. West of the crosscut, the lode was followed for 320 feet at which point the unaltered rock seen in the 60 fm. Level was encountered and the lode became quite barren.

The 90 fm. Level has been driven forward 600 feet westwards along the barytes vein from where the level to Driggith leaves the vein. A very fine pocket of barytes was found beneath the 70 fm. Level workings and a rise has been driven up through for access and ventilation. Some very large vughs were found in this ore shoot, and also there was a very flat place in the vein, about half way up towards the 70 fm. Level. Parts of the vughs were lined with beautiful crystal clear barytes up to two feet thick but on the whole they did not contain a great deal of ore and were also difficult to work. About 1,000 tons of very high grade barytes was raised from the stopes, but a considerable quantity still remains. It was not possible to produce from it quickly enough to make the work profitable.

The last part of the 90 fm. Level west followed a strong leader of barytes which, at times, swelled to nine inches thick. There was no quartz or dolomite, in complete contrast to the 60 fm. Level which, immediately above the point reached, contains a large quantity of both. Near the face a flat, easterly dipping ENE/WSW vein crosses but does not throw the vein being followed. A crosscut was put out SW to see whether there was a stronger barytes vein to the south but this was not so; the intention was to continue the level westwards in the hope of striking a further ore shoot.

Haulage from the 90 fm. Level was by Electric Battery Loco which drew trains of six cars. The tramway continued from the level mouth along the hillside to the mouth of the Sandbeds 90 fm. Level where there was a hopper from which lorries transported the ore over the fell to the mill at Potts Gll. The total production from this mine has been about 35,000 tons of dressed barytes from 3.5 acres of stopes and is equal to a solid rib of barytes 2′ 2″ thick and has been by far the richest mine of the group.

Helvellyn Mining Field

THE Helvellyn area centres on Helvellyn (3,118 ft.) and the surrounding fells. It covers some 40 square miles of rugged country. Mining dates back at least to the Elizabethans but was conducted on a very modest scale until the 19th century. The most sought mineral was galena and the largest production came from the Greenside Mine, where operations were almost continuous for 150 years, finally ceasing in 1962.

The area has been the chief silver lead producer of the Lake District, the output being raised from veins occurring in the middle strata of the Volcanic Series of Borrowdale. There has been but little production of copper, for although several veins have been tried only small deposits of ore were found. Some worthwhile haematite was found near Grasmere, but the heavy transport cost to the smelter, coupled with the then low price for the ore made the mines uneconomic; they closed through these factors rather than through the poverty of the deposits.

Lead veins predominate in the area; many have been found, but few have opened into profitable mines and only Greenside and Hartsop Hall Mines have been worked in this century. Had it been possible to resurrect some of the others they may have proved worthwhile, but the opposition to mining on amenity grounds is now so strong there is little likelihood of any mining being done in this region in the foreseeable future. The area was well-prospected by the ancients and it is doubtful if much of value has escaped them.

The Greenside Mine is not very ancient and is not thought to have been discovered at the beginning of the Elizabethan era, for the first lead mine worked by them in this area was at Grasmere. The Hartsop Hall Mine is of great antiquity and is thought to be Elizabethan, or at least early 17th century. The lead ore there was consistently high in silver and this no doubt would be the chief attraction. Although the Greenside vein was much richer for lead, the silver content of the galena was only one third of Hartsop Hall ore. The only other working which can definitely be ascribed to the Elizabethans is a small copper trial at Thirlspot which, like the Grasmere mine, has never been reopened by later generations.

Early in the 19th century a lead vein was found on Helvellyn high above Wythburn. A syndicate and later a company were formed to explore it and open out a mine, for it was hoped that here would be found another Greenside.

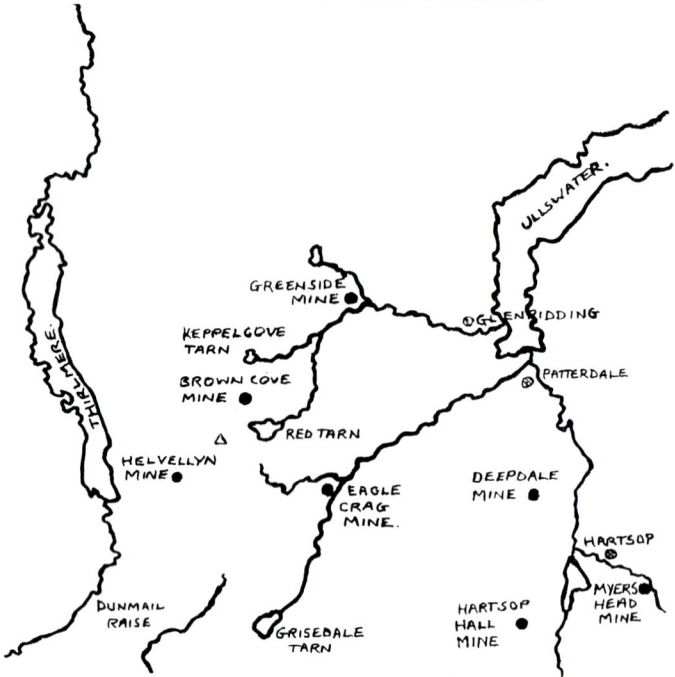

The Helvellyn mining field.

The old Greenside Mining Company was formed about 1822 and gradually interest in the other Patterdale properties waned as it became more and more obvious that there was at Greenside a bonanza deposit. Adventurers put all their resources into this undertaking to make it one of the best lead mines in England.

The bursting of the Kepplecove Dam in 1927 was a major factor in the final decline of the old Greenside Mining Company and although the company was reformed, and capital provided for new equipment, the slump in the price of lead brought the mine to a stop in 1934. The last Greenside Company was formed in 1936 and after spending £250,000 the owners were eventually able to get the mine onto a profitable basis. This state of affairs continued until 1962, when all the worthwhile ore was exhausted. So lead mining in the district came to an unhappy end.

Greenside Mine.

GREENSIDE mountain (2,590 feet) is at the head of Glencoynedale. The Cumberland and Westmorland county boundary runs along the summit. The famous Greenside lead vein runs nearly north and south through the eastern shoulder of the mountain, with its highest

76

point about 2,400 ft. Of the mile of vein explored most of the productive ground lies in Westmorland, as do all the mine buildings, and so the mine is registered as in that county. Greenside was the only lead mine to have been worked almost continuously from its beginning to exhaustion, for in its lifetime it was rarely closed down for more than a year at a time, and it was never abandoned until it was completely worked out in 1962.

The actual date of discovery is not known but was probably about the middle of the 17th century. By 1690 it was in the hands of a party of Dutch Adventurers who are credited with driving the Top and Middle Levels and to have had a dressing plant at the side of Sticks Gill beck. The dressed ore was carried on horseback over Sticks Pass to Stoneycroft Smelter, near Keswick, a distance of about 10 miles. In most winters operations at the mines were at a standstill for two or three months because of the high altitude of the works.

Near the end of the 18th century the Greenside Mining Syndicate was formed; they most probably drove Gilgowers Level and also began the High Horse Level from beside the Washings Dam at about 1,850 ft. It was the lowest point at which the vein could be directly driven upon where the outcrop crosses the plateau at the foot of the upper mountains. The old Greenside Mining Company was formed in 1822 and the mine was carried on with great vigour. At one time it was the most profitable mining venture in the North of England and a 64th share in 1827 was worth £100. By 1837 it was worth £1,000 and it is said that a 100% dividend was once paid out for a three months' period.

By the start of the 19th century an up-to-date smelter had been erected at Alston Moor by the London Lead Company, and after about 1820 the lead concentrates were carted there in preference to the pack-horse trail to Keswick. While the High Horse Level was being driven a crushing and washing mill was erected on the site of the original washings. This new plant had three sets of crushing rollers with trommel screens and mechanical jiggers followed by buddles for recovery of the fines, and the whole was propelled by a waterwheel using water from the nearby man made Top Dam tarn. A cart road had been constructed from Glenridding up the side of the valley to the foot of Lucy Tongue Gill, over which a stone-arched bridge was built (it is still in use today). The road thence zig-zagged up the steep mountainside past the Haystacks to the plateau and the mines. The carts carrying the dressed lead ore to Alston were so numerous that the local authorities insisted that each fifth cart should have its wheels set at a wider gauge; then the ruts in the road would get levelled out automatically!

The bothies, or "shops", stood at the High Horse Level. Here many of the miners from as far away as Keswick and Penrith stayed. In those days the village of Glenridding barely existed, for even by the time the first Ordnance maps appeared in 1860 it was a much less place than Hartsop. The Mining Company had by then built their first two rows of cottages in the lower part of the village and for some unknown reason had also built a row of cottages at Seldom Seen in Glencoynedale. Soon afterwards the houses called the Rakes were built on the hillside, on the way to the mines, and eventually there had risen Stybarrow

Terrace, consisting of two rows of houses in lower Glenridding. Halton Terrace, also of two rows, was built above the mine road but nearer the village than the Rakes, and eventually the Company owned 50 cottages.

By the time the High Horse Level became fully developed a new lower flank level had been commenced from the side of Lucy Tongue Gill. It was 35 fathoms below it and called Low Horse Level. During the 1830s the smelter was built at the foot of Lucy Tongue Gill and a chimney was constructed up the hillside terminating in a stack on the Stang, about one mile away and over 1,000 feet higher. This chimney was mainly a stone-arched cutting following bedrock, or cut into it, and therefore it not only formed the chimney but also acted as a colossal crosscut, covering every possible position of the probable southward extension of the Greenside vein.

The dressing plant to serve the Low Level was built just above and to the south of the smelting mill and was fed by a self-acting tramway down the mountainside. It became obvious about 1851 that the ground above the Low Level was beginning to work out and to win the ore below it a shaft and possibly a far deeper adit from lower down the gill were needed. The Company began sinking the Willie Shaft in the Low Level, about midway between the main oreshoots, and the first level off it was at 36 fathoms. At the same time the Lucy Tongue Level was commenced just above the smelt mill at about 1,100 ft. and 82 fathoms below Low Level.

In 1853, an Alston Moor mining engineer named George C. Cain was called in to make a further survey and he then laid out the proposed route for the new adit. The course of this level is astonishing, bending almost in a half circle with the straight line distance between the mouth and the underground shaft barely half of the total distance driven. However the course taken was quite deliberately planned to take advantage of driving along a soft vein instead of going straight through the harder country rock. The vein was cut in 1862 but the level was not completed until 1868 and was by then 1700 yards in length.

The tunnel was kept going 24 hours per day, with two men hand boring on each of the three shifts. They moved 48 yards per annum in hard rock but by following the softer veins they were able to cut 120 yards per annum. Ventilation was by a waterblast, using water taken into a pipeline high up the gill, and the air from the blast was carried along the level in square wooden boxes. This means sufficed to enable the eventual connection with the Low Level workings to be made, and one can visualise the state of things before the breakthrough, with a thick black smoke from the constant blasting, to say nothing of that from the tallow candles. It is said that fathers started the level and their sons completed it, but when the final holing was made the directors stood these men a grand dinner at the Ullswater Hotel as a mark of appreciation.

The Willie Shaft had meanwhile been sunk a further 12 fathoms and another level was driven out along the vein. All this ore and waste was drawn up by a hydraulic engine and a similar engine worked the pump. Water for driving these machines was stored in the far northern workings of the High Horse Level, and parts of these old pipes and dams still

Smith Shaft Top in the Lucy Level, Greenside Mine.

**Blacksmiths at Greenside
over 60 years ago.**

79

remain. Willie Shaft was sunk through to the Lucy Level about 1882. A shaft was begun from the southern part of the workings on this level in 1862, soon after the vein was cut. The connection from the Lucy to the Willie Shaft 48 Fathoms Level was made through Brookes Sump, which was sunk the 34 fathoms in three lifts, with the bottom section being partly risen from the Lucy.

The Lucy Shaft was eventually sunk to the 60 fathoms level, the power for winding and pumping being supplied by hydraulic engines. Water came from Low Level in a pipeline down through the workings. The name Willie Shaft was from William Glenwright, who was the first winding engine driver, and when there was a rapid thaw after a heavy snowfall Willie is said to have sent word down to the fillers, who happened to be among some rich ore at the time, "not to load so full as how could they expect him to lift the bogies with nothing but snow broth to work with".

Captain William Henry Borlase became manager in 1890 and he persuaded the company that in the face of a falling lead market the only way to survive was to modernise the whole of the mining machinery and to put in a hydro-electric plant to supply power for winding and pumping and driving a new compressor set up in a chamber off Warsops crosscut. An electric loco was also installed in the Lucy and could easily haul 18 mine cars, so doing away with seven horses and their drivers. This machine had trolley arms which picked up the current from two tightly stretched copper conductors set in the level roof, one above each rail. The motor was 30 hp; the machine was controlled similar to a tramcar, variations in speed being accomplished by passing the current through large resistances.

Instead of continuing the Lucy Shaft, a new one called Smiths, was started from a 25 fathoms eastern extension of Warsops Crosscut, and passed through the vein about 45 fathoms down. The first level off was the 40 Fathoms Level, which connected to the same level off Lucy Shaft. The original winding engine on Smiths shaft was the first electric winder installed in a metal mine in the United Kingdom. The Lucy loco was also a first and gave nearly 40 years of service. Further improvements were made to the dressing plant and the whole operation was once again put on a sound footing, a state of affairs which continued very successfully until the 1914-18 war.

It then became uneconomical for various reasons to continue to smelt, and the lead ore was carried in wagons drawn by a steam traction engine (the "Helvellyn") to Troutbeck Station, as had been the pig lead. Before the traction engine era the string of horses and carts which went between the mines and Alston with the dressed ore carried the pig lead to Penrith Station and brought back coal, lime and other materials required for the enterprise. When the Penrith Keswick Railway was built about 1867 this traffic was switched to Troutbeck, a saving of at least six miles. The dressed ore went by rail to Messrs. Walkers Parkers Smelter at Newcastle on Tyne.

In the 1870s a great storm partially burst the Top Dam; the flood down Lucy Tongue Gill carried away part of the silver-refining house and with it a 1,000 ounce plate of silver. It was never recovered, though a gang of men spent several weeks searching through the debris in the beck. Long afterwards, looking the beck for it was a popular pastime after a flood. It was found and spirited away, thought some of the old-timers, but if this was so the truth never came out.

One Sunday in 1862 a fall of ground took place above the High Horse Level. Some 120,000 tons of rock were contained in a pendant-like piece of ground formed as the vein being followed upwards split into two parts and was left unsupported as each branch had been extracted. This huge mass crushed all before it, fortunately when the mine was idle. Examination of the workings afterwards showed that the pendant, which had been considered to be barren rock, had another vein running through it. It took several miners a number of years to recover most of it. The Smith shaft was not continued below the 90 Fathoms Level, but an underlay shaft called Murrays was sunk a little way to the south from the 90 Fathoms, to the 105 and 120 Fathoms Levels.

During the 1914-18 war most of the young miners were called away to the Forces; production fell away and there was little development work. After the war, efforts were made to catch up, and by 1925 the workings had reached 135 fathoms below the Lucy. On 29 October, 1927, a cloudburst washed away part of the Kepplecove Dam embankment and a great flood swept down the mines valley to the lake, damaging houses and shops near Glenridding Bridge. There was no loss of life. The Mining Company could hardly be blamed for it but several people pressed for considerable compensation. Rather than fight a court action, the company paid up. The disaster nearly closed the mines but the company was reorganised and several thousand pounds of new capital provided to repair the damage and get the mine active again.

A water-race was made along the fellside from the mines to Miresbeck in 1928. A pipeline was laid down the fell to a new power plant near Rattlebeck. The price of lead was very low by 1930 and gradually the mines ground to a complete halt in 1934. Pumping was continued and repairs to the shafts and levels maintained. In 1936 the old company was taken over by the Basinghall Mining Syndicate Ltd., of London, and work began on a scale larger than ever before. Almost at once the national grid electricity was brought in from Pooley Bridge and for the first time since the mines began they were no longer dependent on the power raised from their tarns and streams. A fine new dressing plant, incorporating the very latest crushing and dressing machinery, and capable of treating 250 tons of crude ore per day, was erected on the site of the 1871 mill. The Murray Shaft was sunk from the 150 Fathom Level to the 175 and 200 Fathoms Levels and much lateral development work was done.

The new process was not economic and the plant was reorganised to incorporate the best feature of the old system. From then on the operation became a financial success. During the 1939-45 war a shaft was sunk from the 175 Fathom Level to the 200 and the 217 Fathom Levels; this work revealed the Skiddaw Slate and heralded the end of the mine. There was a disastrous fire in the shaft in 1952 and four men lost their lives, being overcome by fumes. After some months the shaft was repaired and the lower levels reopened and work continued steadily until 1959. The Atomic Energy Commission then took over the mine in order to conduct a series of seismic test explosions underground. Two men entered a working which apparently contained poisonous gas and were fatally overcome.

In 1960 the mining of the last ore reserves was resumed and by early 1961 all the worthwhile ore had been won. The mine closed for good in 1962. Thus ended one of the most famous British lead mines; all too soon no men will be left who actually mined there. The site has been cleared up and all the temporary buildings removed. Entrances to the mine have been obliterated to prevent would-be explorers from venturing into the old and decaying workings, many of which are dangerous. The offices, stores and part of the old smelt mill, being well-built of local stone and slate, are now used for mountaineering huts. The miners' hostel is a youth hostel.

The total output from the mine is not exactly known, production previous to the original Greenside Mining Co. is unrecorded. From 1835 to 1961 some 200,000 tons of lead concentrates were obtained, and so in all the total cannot be much below 250,000 tons of lead concentrates, from around 3,000.000 tons of crude ore treated. The overall profit to the various companies involved was about £400,000, the bulk of this being made in the 19th century.

The geology of the Greenside Mine area is fairly dealt with by Dr. David Gough in *Economic Geology*, Vol. 60., No. 7., November 1965. The following description is in more general terms.

The Greenside Vein is associated with a fault fissure running NNE-SSW and dipping steeply east through the Volcanic Series of Borrowdale; it has a traced length of about a mile though the mineralised part of it does not exceed half of this distance. In the northern part of the mine the vein is in association with a quartz porphyry dike, through which the vein passes, and at this point an E-W crosscourse dipping steeply north, and known as the North Crossback, shifts the lead vein slightly to the west and appears to have a direct bearing on the productiveness of the latter.

The northern limit of the vein has not been reached, but the 500 feet driven forward from the most northern orebody does not look particularly encouraging in any of the four levels driven to this point. The southern limit of the vein is marked by a very strong E-W crosscourse known as the Clay Vein. The Greenside vein does not exist south of the Clay Vein, a fact which seems hard to believe when you consider its great size and strength. The original miners clearly did not believe it could possibly die out and the course and construction of the smelter chimney was deliberately planned to cover a wide range of dislocation. The negative results from this work proved to them that it did actually die out, and as no underground trial was made in this direction by the old company it shows that they were quite satisfied.

In the Greenside vein, ore-bearing ground is filled with broken country rock, sometimes a great deal of quartz, and in the higher levels a considerable amount of barytes. The only economic mineral is galena which occurs in ribs, strings and blotches throughout the mass. In the ground below the Low Level the barytes disappeared and it would

G.E.C. locomotive, the first electric trolly, photographed in 1891.

Filling a wagon from the hoppers, Greenside Mine, 1900.

83

Fungus which formed on the walls of a level in Greenside
(a photograph of about 1900).

Two partnerships of miners having their ''bait'' in a ''driving'',
Greenside Mine, about 1900.

seem that there was an increase in the galena. Some of the quartz had a peculiar brown marking and was then extremely tough. It was known by the miners as "throstle breasted ground" and was reckoned the most difficult to break. Between the oreshoots the vein was filled with light-coloured soft rock called "chunk", and there were many false walls. The only true one had a thin black marker or leader.

In the Low Level period of working a great deal of exploration was done on this horizon. The level starts from beside Lucy Tongue beck and is driven along a small irregular barren vein with a general NW bearing leading to the Greenside vein, which it cuts near the southern end of the south oreshoot and at about the junction with the Clay Vein. A vein running SW from this point, Marshalls vein, has been followed for over 100 fathoms but no lead ore was found. At 75 fathoms, north of the junction, the Johnson Ross crosscourse passes through the Greenside Vein. The dislocation caused by this fault is very slightly to the west but it does not appear to have any influence on the mineral content of the vein. Nothing is known about this fault Eastwards but to the west it continues strongly and above Thirlspot, about two miles away, it contains copper. This was worked in a small way by the Elizabethans.

The southern Greenside ore shoot was discovered by the Low Level, as its outcrop is thickly covered by boulder clay and so it was not seen by the ancients. This ore shoot was of short length but great width and was fairly rich, although the galena always carried less silver than in the rest of the mine. It is doubtful if all the ground stoped on this shoot on the Low Level is shown on the old plans as on the 36 Fathoms Level, immediately below, the shoot has been stoped for 100 fathoms, this being the greatest length of ground worked on it in the 300 fathoms of vertical extent exploited.

The main ore shoot throughout the mine was the one first discovered high up the mountain near to the county boundary, which curiously marks almost exactly the northern limit of the shoot. The first miners drove the Top Level directly along the vein at about 40 fathoms below the summit, and they stoped the vein over the whole 90 fathoms driven right out to surface.

The Middle Level, 20 fathoms lower down the hillside, has been driven to the end of the orebody and entirely stoped out. Gilgowars Level is 25 fathoms below the Middle Level and has been driven directly along the vein for 180 fathoms of which all but 20 fathoms has been stoped up through to Middle Level; the surface above has the appearance of a great quarry hole. It is thought all these workings were started before the Greenside Mining Syndicate was active; this syndicate are credited with starting the High Horse Level, which is about 15 fathoms below Gilgowars Level. The High Horse Level starts from beside Top Dam as a crosscut 40 fathoms in length though it could actually have been started directly on the vein. To do this would have meant a long cutting and a timbered or stone-arched drive through the thick boulderclay, so by starting where they did the miners avoided this. The rock is almost out to day at this point.

The "Great Crush" took place in this level and removed the floor of Gilgowars Level. The main ore shoot has been stoped out for 160 fathoms in length but the level has been continued northwards for a further 117 fathoms. At the face a short rise connects with Glencoyne Level, which is the only working in Glencoynedale, starting as a crosscut 40 fathoms in length to the vein. Immediately above this point, on the surface, is an ancient trial known as Dukes Sump, the most northerly surface point at which lead has been found in Greenside vein. Glencoyne Level is driven along the vein south for 60 fathoms, to the connecting rise, and northwards from the crosscut for 80 fathoms. It was finally abandoned in 1873, for no lead ore was found in any of this work.

Low Level crosscut, which follows a small barren vein, is 275 fathoms in length and 35 fathoms below High Horse Level. The total length driven along the vein is 670 fathoms and the level extends 70 fathoms further north than the Glencoyne level, being given up in 1877. Old plans show that some 230 fathoms in length on the main shoots are stoped out, but it is believed that a further 40 or 50 fathoms of length has been stoped but not recorded. Almost immediately below Duke's Sump a small patch of ore was found and stoped out for a length of 25 fathoms to a maximum height of 15 fathoms above the level. A sump was put down and the ground stoped out to a depth of 10 fathoms below the level. This work was done so long ago, and had been inaccessible for so long, that even in 1925 there was only one man left alive who had actually been there.

In 1937 the Basinghall Mining Syndicate cleaned out the Glencoyne Level and put a ladderway down the rise to the Horse Level. At about 100 fathoms along this level there had been a sump sunk in dead ground down to the Low Level. A new ladderway was fixed down it in the hope that it would then be possible to reach the old stope but large falls had completely blocked the level a little way northwards. To the south, much of the vast old stopes had collapsed and, in any case, this area had been totally collapsed north of Willie Shaft for more than 50 years. There was no point in trying to clear a way through in this direction.

When the decision was made to drive the Lucy Tongue Level, the intention was to follow the Lucy vein up to the Clay Vein and, after passing through it, to crosscut north-west to reach Greenside Vein. Work was started following the Lucy Vein but this was barren, similar to the vein on which Low Level had been driven. After Mr. Cain made his survey of the project in 1853 he predicted closely where the Clay Vein would be encountered and suggested that the level should then be turned west along this fault, which would lead the miners towards the Greenside Vein. After some 200 fathoms had been driven along the Clay Vein the level was again turned off north-westwards to cut the Greenside Vein about the centre of the South Ore Shoot. A great deal of the distance driven along the Clay Vein is not self-supporting and the ground was carefully stone-arched.

The Lucy Vein appears to pass through the Clay Vein and is therefore still younger. Generally it is narrow but contains a good deal of soft material, with odd bits of pink barytes and some quartz. A little further driving was done through the Clay Vein at the foot of Low arches in 1929 and though the vein followed was very strong, it is not clear that it was the Lucy Vein. No metal values of any sort were seen. Clay from this working was used for stemming and the miners called in when going off shift and filled themselves a car, which was then sent down to their working place. The Lucy Horizon was not a particularly good one for ore and must have been disappointing for the Adventurers who spent much money and effort in creating the adit. They were rewarded by the great improvement of the vein in the 40 Fathoms Level, which encouraged them to drive the Lucy a little way beyond the limit of the oreground worked in the 48 Fathom Level above, and the Alma Shoot was discovered. This extended for some 35 fathoms and was followed upwards with an intermediate level off at 25 fathoms.

The 20 Fathoms Level off the Lucy Shaft was not extended beyond the South Shoot, but the 40 Fathom Level was driven full length of the known ore zones a total distance of 325 fathoms. At about 90 fathoms north of the shaft a short but very rich new shoot was discovered and called the Middle Bunch. It is said that here was found the most solid galena ever seen in the Greenside Vein, but in depth the shoot was erratic, being sometimes rich and often very poor. To ventilate the 40 Fathom Level, Hicks Sump was sunk at a point 200 fathoms north of the Lucy Shaft. A steady flow of water was encountered which is said to have amounted to 16 barrels per hour. As all the ore and water was in those days wound from sumps by hand this was serious and to speed up the work the sump was taken down in three offset stages of 13 fathoms each. When the bottom section was being made three sets of windlass or jackroll men were employed on each of the three shifts. The total depth of the sump was reached ahead of the arrival of the 40 Fathom Level, and so it was allowed to fill with water. The 40 Fathom Level then proceeded past the supposed position of the sump without holding to it and the manager, when inspecting the face, complained to the miners about a bump left on the side of the level a little way back from the face. They agreed to put a hole in it and blast it off, but when they were boring the hole the drill steel went through into the sump and the water pressure pushed it out of the man's hand. The resulting rush of water very nearly overwhelmed the pump.

Capt. Borlase became manager about this time and he sank Smith's Shaft from an eastern extension of Warsop's Crosscut. This shaft was sunk with rock drills and, being off the vein, was dry and soon reached the 60 Fathoms Level. The price of lead was low, only £9 per ton for the pig lead, but the company were fortunate to find the richest ground so far worked. The Alma and a large area at Hick's Sump were left unworked until such time as the lead price improved. Smith's Shaft was discontinued at the 90 Fathom Level and a Skip shaft, following the footwall of the vein, was sunk to the 120 Fathom Level. This system was not very successful because of the double handling involved and a new shaft called Murray's was begun. This followed the average underlay of the vein but was equipped with cages so that ore cars, men and ponies could be lifted.

Between the 120 and 105 Fathom Levels, the South Ore Shoot always referred to as "The wide ground") was up to 40 feet in width. There was an unfortunate accident in which one of the miners was killed by a fall of ground. When the 120 Fathom Level was being driven forward from the North Crossback in 1929 a strong flow of water was cut and it took two weeks of pumping, night and day, before it ran off, and clearly a vast area of vein from somewhere ahead of the known ore limits had been drained. The price of lead was dropping so low that despite every economy the mine was unable to pay its way and it was not possible to drive the level far enough forward to try to find this possible new orebody. When the Basinghall Mining Syndicate took over they continued the level and opened into a splendid run of ore 150 fathoms in length, though it did not extend upwards for more than 10 to 15 fathoms. At the end of this shoot a fault called the Wynn cross-course was passed through, with negligible dislocation of the vein, and north of this point there was a further shoot some 50 fathoms in length which continued upwards for 40 fathoms.

This level was continued northwards for some distance but failed to find anything of importance. In view of this discovery on the 120 Fathom Level, the Lucy was restarted and driven forward a total of 300 fathoms past where it had been given up by the old company. No ore of importance was found and it was proved that there was no connection between the little shoot found in Low Level so long ago and the new ore on the 120 Fathom Level. The Murray shaft had been taken to the 200 Fathom Level but the Southern ore shoot, which had been growing progressively poorer for some time, had only 30 fathoms of length worth working; even that was poor, consisting mainly of hard white quartz with the galena scattered throughout.

It was decided that it would be more economical to sink a new vertical shaft towards the north end of the 175 Fathom Level, and this was begun in 1942. It was sunk to the 217 with levels off at 200 and 217 Fathoms. Just above the 217 Fathom Level the Skiddaw Slates were encountered, and although the vein continued it did not contain any ore. All the worthwhile ground above the 217 Level in the old and new North shoots was stoped out, and a short rich section was followed down to the 237 Fathom Level, where the flattening of the slate pinched the ore out. The beds of Skiddaw Slate encountered appear particularly unfavourable for ore and although several boreholes were put down no ore of importance was found. The so-called Skiddaw sandstones and the more massive slate beds are more favourable horizons in the Keswick area, and it could be that there could still be a large amount of ore much deeper down. This depends on whether these beds are there. The slate so far seen is high up in the series, a point which would be difficult to prove or disprove in view of the contorted nature of these rocks.

In the lower levels of the mine the length of ore ground was not much different to that in Low Level but was spread over a greater length, with long stretches of barren ground between the shoots. The average grade of all the ground stoped is thought to have been about 8% galena, but the width worked varied considerably ranging from six feet up to as much as 50 feet. The veinstuff and country rock were tough and strong and it was possible to work the mine on the flatback shrinkage system of stoping, a method used almost entirely from start to finish without having to resort to support of any kind.

For many years the only way in and out of Greenside was by way of

Lucy Level and Smith Shaft. This was due to the collapse of the timber-work near the top of Willie Shaft, which by their own weight completely stripped it from top to bottom; it was never repaired. Low Level fell into disrepair and was abandoned. When the authorities pressed for a second emergency way out to be provided it was made through Alma to Low Level and up the ladderway to Glencoyne.

Capt. Borlase built the Top Power House, near where Red Tarn and Kepplecove becks join together, in 1890. A deep cut was made into Red Tarn and a low earth dam constructed. At the same time a water leat was made from near Kepplecove Tarn outflow at about 1,750 ft., along the fell breast at the foot of Catstye Cam to Red Tarn beck and along the hillside to the intake tank for the turbine pipeline. This gave a vertical head of 400 feet and there was sufficient water to generate about 200 horsepower. Power was at first generated at 600 volts D.C., being taken on overhead line to Low Level. Here it was conducted along the level in two separate lead covered cables which were suspended down Willie Shaft to the Lucy Level. About 1912 more power was required and the original plant was replaced by a modern Gilkes impulse turbine direct coupled to an alternator, giving 2,200 volts A.C., and this power was taken by overhead line to the Lucy Level mouth; a threecore armoured cable carried it up the level to Warsop Crosscut. Here a transformer and rotary converter changed the current to 600 volts D.C. for the various machines. A 150 h.p. Impulse turbine alternator set was installed alongside the Dressing plant in 1924. It used water from Top Dam, on about 500 feet head, and was chiefly used to supply power for a 500 cu. ft. min. Alley McLellan Compressor set in a chamber off Warsop Crosscut.

In 1928 the power plant was again reorganised and improved by installing a new turbine generating set near Rattlebeck with overhead line to Lucy Level mouth. The water for this plant was collected at the mines on a level with the Lucy and conveyed along the breast of Blea Cove in a wooden launder about a mile in length. The head at the turbine was 365 feet and it developed about 300 h.p.

When Basinghall Company took over they scrapped all the old milling plant and built a modern floatation mill capable of treating 250 tons of crude ore per day. The crushing section was located just below the Lucy Level mouth and consisted of a large Blake type stonebreaker and a Symons cone-crusher working in closed circuit with a Symons vibrating screen. The finely-crushed ore was carried up a long con-veyor to a large steel bin from which the various dressing machines were supplied, being located in the former Table house of the old plant. Each machine was driven by an electric motor and power came either from the company's own powerhouses or from the national grid, depending on the water situation. The gravel tailings from the plant were run into a large wooden hopper and over the years a good market for this material was built up and so most of this waste was sold. The fines which constituted the major portion of the tailings were pumped to a disposal dam built on top of the old Lucy waste dump. The layout of the system made it semi-automatic in building up the outer perimeter, and the dam was well-constructed, a systematic drainage network making sure that as it built up the lower portion was properly drained.

When the mine ceased production, the machinery was dismantled and removed from inside the mine as well as the mill and workshops. The site was tidied up as much as possible to its original state. Getting vegetation to grow on the tailings heap proved difficult because of the lack of organic material in the sand, and experiments with various types of pampas and other grasses were carried out over a number of years. A large quantity of sludge was brought from the sewerage works at Penrith and when this was spread over the dump it provided the necessary start for vegetation; ultimately it is hoped that nature will so camouflage the scene that few will recognise that there was once a thriving and successful enterprise.

Eagle Crag Mine.

THE Eagle Crag Mine lies about 2½ miles south-west of Patterdale village on the west side of Grisedale valley and into the eastern slopes of Helvellyn. When the mine was opened is not known but some of the workings are very ancient. Much of the work was done in the 19th century but it is doubtful if it was ever a profitable mine. No original plans or sections are known to exist and although some of the levels have been closed up for a long time it is obvious that the workings are much more extensive than the accessible parts show. About 100 years ago the mines were taken up by the Patterdale Mining Co., a syndicate of men who were mostly the main shareholders in the Greenside Mining Company, with Mr. George Head-Head as chairman and Richard Grenfell as manager of the mines.

With the sudden death of Mr. Grenfell in 1877 coupled with the lack of mining success, the undertaking closed and nothing has been done since. A pocket of lead ore discovered in 1837 yielded 30 tons of solid galena, no doubt a rich prize in those days, when most of the miners were self-employed and paid by the mining company for the lead ore actually produced.

The main vein runs E—W and dips slightly south up the steep face of Eagle Crag. It is thought to continue westwards for about two miles to the Wythburn Mine. On the whole it is a narrow but persistent fissure, rarely exceeding 18 inches in width. The galena, the principal mineral, runs in a rib on one of the walls and the rest of the vein is composed of broken country rock and some open, crumbly quartz. The eastern end of the known part of the vein terminates at a crosscourse in the Grisedale beck and a quarter mile further east a crosscut has been driven into the base of St. Sunday Crag at about 850 ft., but has long been closed at the mouth. The size of the spoil heap indicates that it has been driven a considerable distance apparently without success, as no sign of mineral is to be seen. The vein must not have been found. Many years ago a similar-looking east-west vein, roughly on the line of Eagle Crag vein, was discovered on the fellside behind Beck Stones farmhouse in Patterdale and though it contains a little galena at grass no attempt has been made to open it out. This point is about 2½ miles east of Eagle Crag Mine.

HELVELLYN MINING FIELD

Near the end of the 19th century two brothers named Watson, who resided in Deepdale and were employed at Greenside Mine, found a vein carrying galena at the surface on St. Sunday Crag. They obtained a piece of ore showing good galena and concealed the spot with sods. There was thick mist at the time but they were so familiar with the ground that they were sure they could find the place again. They took the piece of ore to Captain Borlase at Greenside and he was impressed and suggested that they should take him to see the place when he could make an examination with a view to deciding on the best way to make a trial. They could not locate the spot and spent many days in abortive searching. Over the years many people, including my father and myself, searched the area but no trace has ever been found and had it not been for the piece of ore shown to Capt. Borlase it would have been held to be a hoax. Watson's Lost Vein remains a mystery.

In all, seven levels or adits have been driven along the course of Eagle Crag Vein. The lowest, or No. 1 Level, starts direct on the vein about 60 feet above the level of the beck which is at 780 ft. O.D. at this point. A little piece of ground has been stoped out down to the level of the beck between the level mouth and the crosscourse but the vein seems to have been very poor. The No. 1 Level extends at least 20 fathoms west, at which point a shallow rise has been put up to surface. The extent of the level beyond this point is unknown and but little ore seems to have been found.

No. 2 Level is 20 fathoms above the No. 1 and again starts direct on the outcrop in solid rock. This was the last level to be worked, disappointingly, driven a total of 150 fathoms, of which only the first few fathoms were reckoned worth-while to stope out. The vein was shifted slightly south by a small crosscourse at about 25 fathoms in from the level mouth, and at 100 fathoms a rise connects with the No. 3 Level. Some good specimens of Hydrozincite were obtained from this level; they formed on the floor over the years since work ceased. No. 3 Level is 15 fathoms above the No. 2 and starts near the foot of the steep crag and has been the main production level of the last company.

Patterdale Mining Company rarely had more than four miners at constant work and sold a few parcels of lead ore obtained by hand-dressing. There was no mechanical washing plant. Had the development work succeeded, such a plant would have been provided, for there was abundant water readily available for propelling waterwheels. A good stone house was built near No. 3 Level, apparently for use as an office and a shop at which some of the men could stay. There was also a smithy, a powder house, ore-sorting and hand-washing floors.

The total extent of No. 3 Level is not known but it is about 75 fathoms in to the connecting rise from No. 2 Level. This level is older than the 1870 operations and from the size of the tailings dump it is apparent that there has been a lot of work done. No doubt several hundred tons of galena were raised, and the stoping must be considerable.

No. 4 Level, driven about 40 fathoms above No. 3, has long been closed and there is no idea of its extent. Some of the stopes above No. 3 Level will probably connect with No. 4 Level. No. 5 Level, at about 1,650 ft. has been driven at about 25 fathoms below the top of the crag and for at least 125 fathoms westwards. Stopes totalling 61 fathoms in length are to be seen at the surface, and these were worked upwards from this level and terminated against a north-south crosscourse called the Clay Vein. This crosscourse is slightly mineralised in parts, and north of Nethermost Cove beck a level has been driven upon it and some galena found though apparently not in workable quantity.

At No. 5 Level there have been hand-dressing floors similar to those at No. 3. I suspect that they are very much older. No. 6 Level has been driven as a shallow level eastwards towards the top of Eagle Crag and so cannot extend far. No. 7 Level is a crosscut driven south to the vein at about seven fathoms below No. 6 and is connected up to No. 5 level by stopes and rises.

The Eagle Crag Vein west of the Clay Vein has not been tried and the continuation was not found, although some miners thought it to be the east-west vein at Wythburn Mine. The northern level on the Clay Vein was reopened and explored in 1874 with negative results. The narrow Eagle Crag vein is nevertheless extremely persistent and the galena occurs chiefly as a solid rib which made the ore relatively easy to dress, a factor which accounts for the large amount of work done by the ancients.

The long distance driven in No. 2 Level with such poor results leads one to ponder whether good ore ground occurs further ahead in the No. 3 Level workings and if this was the reason for the start of the level. If so, how much further driving remains to be done to reach it?

The hand-dressed galena ran about 75% lead and carried 16 ozs. of silver per ton. It is very doubtful if permission could now be obtained to reopen the mine.

Hartsop Mine (often called Hartsop Hall Mine).

THIS ancient mine lies about a quarter mile west of Hartsop Hall and 2½ miles south of Patterdale village. It is near the foot of a steep mountain but, unlike most mines, the entrance is in a wood. The vein runs NE-SW diagonally along the mountainside and dips steeply west wards at varying angles up to 25 degrees from vertical. The country rock is Borrowdale Volcanic ashes, through which run numerous quartz strings, but it does not appear clear whether these have any bearing on the mineral content of the vein.

Much of the workings are near the surface and stretch along the outcrop for over a quarter mile. They show the vein to be not more than three feet in average width and to be composed of quartz and brecciated country rock with a good deal of barytes, the only mineral of value being argentiferous galena. Trifling amounts of copper and blende have been seen, the latter in the deepest workings only.

From the middle orebody section of the mine some small but perfect cubes of pale straw coloured flourspar have been obtained and are the only specimens of this mineral to have been found in the Borrowdale Volcanics. The late Sir Arthur Russell also obtained specimens of wulfenite from the same oreshoot. The galena is generally finely-scattered through the veinstuff, being rarely in solid ribs, and is therefore difficult

No. 1 Level, Hartsop Mine, in 1931.

to dress by gravity means. The silver content averages 30 ozs. per ton of galena when dressed to 80% lead. A little gold also occurs in the galena but not in sufficient quantity to be paid for by the smelters although, no doubt, they would eventually recover it when refining the silver.

The ore shoots are separated by stretches of barren ground very similar in character to the Greenside Vein. One or two branches are thrown off to the north-west but do not bear ore away from the junction with the main vein. At the southern end of the workings the vein comes up against an east-west crosscourse, which hades north; it has not been seen to the south of this. The surface here is covered with thick glacial till full of huge boulders and the height on the line of the vein above the valley floor is rapidly diminishing. It is little wonder that no further surface crosscut has been driven to look for the probable continuation of the vein. Why no trial was made through the crosscourse in the lowest or No. 4 Level is not apparent.

MINING IN THE LAKE COUNTIES

Many years ago my father discovered a vein, which we called Thomas Vein, lying about 100 feet west of the Hartsop Hall vein. It showed up in a small stream after a great flood and appears to run NE-SW and dips to the west, carrying up to 18 inches of barytes and a little quartz. Although some digging was done there was no sign of any galena. It is not clear whether this vein is south of the crosscourse, but if it is, then most likely it is the continuation of the main vein. It would require a righthand throw for this to be so, and in Lakeland this is unlikely.

In the lowest crosscut (No. 4), about 12 fathoms along the vein NE of the crosscut, a copper vein is passed through; this vein, which strikes east and west and dips steeply north, is three feet wide and composed chiefly of quartz with a little chalcopyrite of good quality. The lead vein displaces the copper vein some six feet horizontally to the north. The surface sumps or bell pits and the No. 1 crosscut are pregunpowder- and there appears to have been a good deal of ground stoped out from them by the old miners.

About half a mile SW of the mine are the remains of an ancient smelter, the main building of which measures 25 feet by 15 feet, the hearth and chimney being in the centre. This arrangement is similar in construction to one shown by a woodcut in Agricola.

Water from the Hogget Gill was led into a small dam from which a launder had carried it to the waterwheel propelling the bellows. Nearby are the ruins of a beehive-shaped charcoal house and there is a pile of slag, both glassy and metallic, in which a careful search will yield small pieces of metallic lead. In the old buildings were found the oxidised remains of a pile of galena, so decomposed it must have lain there for centuries. The stone to build the works had been quarried nearby and in the walls were several large pieces of hard white quartz which had probably come from the quarry. Some contained a peculiar kind of antimony sulphide, but none could be discovered in situ in the strong quartz vein running nearby. This old smelter must have been part of the original mining operations, late 16th or early 17th century.

The mines were reopened and energetically developed in the 19th century and these operations must have continued for many years. It has not been possible to trace who the Adventurers were, or to find any records. New levels were driven below the earlier ones and a great deal of ore ground was laid open. A water leat over a mile in length was constructed to bring water from Dovedale beck to turn a waterwheel driving the crushing and washing machinery. A smithy, joiner's shop and office were built a little way below the mill. When the mine closed about 1860 the buildings were pulled down and the stones used to build a new barn at Hartsop Hall and also the new part of Brotherswater Hotel. The waterwheel is said to have gone to Ambleside where it was re-erected beside Stock Beck to propel wood-working machinery. The last work done in the mine by this company was the sinking of a sump below the lowest or No. 4 level which showed some four inches of solid galena.

In August, 1888, the Greenside Mining Company had the old workings reopened and the following notes are from a report made to them by John Comer, Underground Captain at Greenside, who was in charge of this work: "The bottom crosscut is 27 fathoms long. Cleared along the vein 37 fathoms total 64 fathoms. No. 2 crosscut is 17 fathoms long. Cleared along the vein Northwards 63 fathoms. Southwards 44 fathoms. The height between the levels is 20 fathoms 3 feet at the sump. The height of No. 2 Level above peg at Farm house (Hartsop Hall) is 47 fathoms 2 ft. Distance to drive from the peg to the vein is 125 fathoms."

Nothing more was apparently done but the fact that Greenside Company made the examination is interesting. Many of the miners who had worked at Hartsop would still be alive and possibly some plans and records were available. They seemed satisfied that most of the oreground above No. 4 Level was worked out but, had they or the old compay realised that all the ground below No. 2 level was intact, things might have been very different.

When the mine was reopened in 1931 the old company's sump was reached in No. 4 Level. When it was being unwatered it was found that the old men had set timbers across about 10 feet below the level and had built up to the sump top with waste. The water had receded below the timbers and they suddenly gave way. It was never cleared out and the ore in the floor has not been seen. In the 70 years or so since this sump had been sunk the stagnant water, on being pumped out, gave off a great deal of sulphuretted hydrogen. A miner collapsed and the others were affected for several hours after returning to the fresh air. Before the timbers gave way the jackroll and old kibbles were seen to be just as they were left; they were all swallowed up and lost.

My father knew one of the old miners who made this sump. He was William Hellen, of Coniston and he was an old man when he told the story of the good vein. This was a reason for our family reopening the mine. In 1931 a Syndicate consisting of my father, Thomas Shaw, his brother Jonathan Myles and myself began to reopen the mine and No. 4 Level was cleared to Hellens Sump after a good deal of work, much of it spilling through old fallen stope fillings. The old men had used wooden rails and these were intact. They were made from elm and were $3\frac{1}{2}$ inches square and pegged down with oak pegs on to oak sleepers. The curves had been made in the joiner's shop and had probably been laid by the joiner himself. A narrow strip of flat iron had been nailed flush with the inside edges so the flanges of the wheels would not bite into the wood. No bogies or wheels were found but this was not surprising. In those days they would be costly.

The No. 2 Level, which is about 25 fathoms above No. 4 and a quarter mile NE was also partly cleared. All the ground broken in this level had been removed by wheelbarrow; at some places ruts had been made in the rock by the constant passage of the iron wheels. No barrows were found, but years ago one was discovered in a loft at Hartsop. It was similar in design to those used in the mines in Agricola's time. Two planks were used to make the sides, being cut out to take the wheel at the front and shaped into the handles at the rear. There were no legs and the body sat direct on the ground. The wheel was of large diameter made of cast iron and hooped similar to a cart wheel.

MINING IN THE LAKE COUNTIES

Some of the old stopes were rarely more than 18 inches wide and had been worked on a kind of shrinkage system in which the richest ore was sorted out after each blast and thrown down a chute which was built up through the accumulating broken material, to be drawn out for the mill. When the stope was completed the rest of the stuff was drawn out for grating; the greater part of the "deads" were removed and so the actual tonnage of ore to the Mill was small but rich.

Above this level a little ore remained to work and when it was broken was tipped down to No. 4 and trammed out to the mill. This was set up on the old site and consisted of crushing rollers, two small jigs and a Diester table, all belt-driven by a 16/20 hp Marshall oil engine. Alternatively the engine drove a small compressor which supplied air for one rock drill. Unfortunately by the time all this work was completed the prices of lead and silver had fallen and it was impossible to carry on. The mine was closed pending a price rise. Work was restarted in 1934 and a crusher, three stamp battery and slime table were added to the mill. Much of the galena was so finely distributed through the ore that recovery was only about 60%.

Some excellent ore was found in Joe Sump in No. 2 Level, and when the level was cleared to the face of the far north stope it was found to be worth working, the venture looking more promising as regular small shipments of dressed ore were made. Lack of capital prevented installation of a more powerful compressor with air hoist and pump which would have enabled much more of the rich ore to be won from below No. 2 Level. In 1940, in conjunction with the late Mr. J. H. Clay of Caldbeck Mines Ltd., Hartsop Mines Limited was formed with a capital of £2,000 and an 85 hp. Crossley Diesel engine and a 250 cu. ft. min. compressor were installed. To increase the throughput of the mill, and to try to improve the recovery a Huntington mill and two more tables were installed. The No. 4 Level was driven forward and intersected the downward continuation of the main ore shoot worked in No. 2 Level. A sample, taken by the Ministry of Supply's mining engineer from the forehead of the level gave 8.26% lead and 2.5 ozs. of silver per ton of crude ore across a width of 34 inches of vein.

At the outbreak of war it was not possible to continue and the operation was shut down in 1942. The ore was not very amenable to gravity concentration but would be ideal for floatation and it was hoped that some arrangement could have been made for the ore to be dressed at Greenside. With the death of the Earl of Lonsdale this part of his estate was taken for death duties and passed to the National Trust; the conditions which they wished to impose as to the scale of working the mine were not practical so the company had to give up, all the plant was cleared and the site left as it was in 1931. Thus ended many years of endeavour and hardship by working miners to try to revive one of Lakeland's oldest lead mines. It was beaten in the end by circumstances which had not been remotely considered when assessing the risks of the venture.

The main details of the workings are as follows: The No. 1 Level is a short crosscut to the vein which it cuts at 5 fathoms below the outcrop. The level extends each way along the vein for some distance and much of the ground above it has been stoped out. All these workings were pre-gunpowder and no doubt made by the original owners. The No. 2 Level starts immediately below No. 1 at about 900 ft. O.D. and is 8 fathoms lower down the hillside. Carved in the rock at the entrance to this level is a "fleur de lys", though what it infers is not known. The crosscut is 17 fathoms to the vein, which has been stoped out for a length of 20 fathoms; with a connection to No. 1. SW of the crosscut the level extends 37 fathoms along the vein and throughout the distance carries a little galena. Some of the ground has been stoped out.

The level ends at the Orepass sump which connects with the No. 4 Level. At 10 fathoms NE of the crosscut is Joe Sump which was sunk 10 fathoms deep by the old mining company. The syndicate stoped out a rich little shoot of ore above this sump and also did some underhand stoping off the NE side of the sump. There is a small quartz crossvein near the sump top which is displaced 6 feet to the north on the west side by the lead vein. The level continues for 38 fathoms along a barren vein after leaving this oreshoot, and then opens out into a narrow rich shoot, some 20 fathoms long which has been stoped up through to some very ancient surface workings. From the NE end of this stope, at 2 fathoms above the main level, an incline drift known as the Birdcage was driven for 12 fathoms; only a little ore was found. The main level came into good ore at 15 fathoms from the end of the stope and this ground was worked upwards for 6 fathoms, with ore still showing in the roof. Above here an old trial crosscut (wood crosscut), shows a little ore in the floor and very likely the ore in the stope would connect with this if it was worked.

No. 3 Level was started from the surface against the crosscourse which marks the southern limit of the workings. It is a shallow level driven NE. along the vein. There are signs of holes for gunpowder but the original level may be very ancient, being widened by the later miners. How far it extends is not exactly known but it definitely goes past the Middle bunch stope, the far end of which is 67 fathoms NE of the crosscourse. All the workable ore above the level has been stoped out to surface and from some of the old surface sumps there are underhand stopes which go down some distance below the level.

No. 4 Level, the adit has been the main working of the old company and more than likely was all driven by them. The crosscut is 27 fathoms to the vein which comes up against the crosscourse at that point. After passing through 20 fathoms of almost barren vein the level has been stoped out to No. 3 level for a length of 48 fathoms. There is then a barren stretch of 45 fathoms up to Hellens Sump which has been sunk where a westerly branch is thrown off and where the Orepass Sump from No. 2 Level comes down. This branch was driven on for 15 fathoms but very little ore was found. The main level continues forward for 25 fathoms, the vein over this distance being stoped out almost to the No. 2 Level. At this point a West branch, which is seen in No. 2 Level near the Orepass Sump, but has not been tried, leaves the main vein and for some reason the old people drove forward for 12 fathoms along it, leaving the proper vein untouched. As with all the branches, the values soon died out and the face of the level is completely dead.

In the long period the mine lay idle, before being reopened in 1931, the timbers holding up the fill material in this large stope gave way. It was decided that rather than try to clear a way through it would be more satisfactory to drive a connecting crosscut from the first branch through into the second one; and this was done in 1940. This crosscut was then continued eastwards and picked up the main vein some 3 fathoms in beyond the second branch. The level was continued 10 fathoms along the vein, which shows every promise of opening into a valuable deposit. This face is standing in ore.

In 1948 a composite sample of the crude ore remaining in the various hoppers in the Nos. 2 and 4 Levels was obtained for Messrs. McKechnie Brothers Ltd. and the assay of this was as follows:

Lead	20.36%
Zinc	3.80%
Barytes	4.40%
Silver	5.53 ozs. per ton of ore.
Gold	0.016 ozs. per ton of ore.

This was the last time my father and I were in the mine, and shortly afterwards both Nos. 2 and 4 Levels were blocked by falls. This vein will someday be reopened and worked.

Myers Head or Low Hartsop Mine.

MYERS Head is situated a little south of Hartsop hamlet at the junction of Pasture beck and Hayeswater Gill. The vein is seen in the beck bottom at this point and continues SSE along the course of Pasture beck for some distance. By the eastern side of the beck are several very ancient pits, now filled up. It could be that the old men obtained ore from them using workings beneath the beck. It is though that these workings were probably made at the same time as the early ones at Hartsop Mine and that these people were responsible for errecting many of the houses in Hartsop hamlet, which had for many years the largest collection of houses in Patterdale.

About 1870 the Patterdale Mining Company with Richard Grenfell as manager, commenced sinking a vertical shaft on the south bank of the stream opposite the junction with Hayeswater Gill. A 30 feet diameter waterwheel was set up by the side of the gill to work a Cornish pump; the troughs carrying water to it were set on massive stone pillars which remain almost intact. The driving rods for the pump were similarly supported across the main beck and at the shaft top an angle bob changed the horizontal motion to vertical to lift the pump rods in the shaft. A wooden headgear and horse whim was set up to draw the kibbles out of the shaft and it was then sunk to the 30 Fathoms Level. Why this depth was attained before driving out to the vein is not clear but it could be that Mr. Grenfell had some information about the extent of the ancient workings and did not wish to break into them.

The vein seen in the beck bottom is composed mainly of stone and quartz; it runs NNE-SSW but southwards it splits into two branches which have been tried from shallow levels on the 700 ft. contour. These levels are little higher than the shaft top but start 100 fathoms from it.

The westerly branch does not seem to have yielded anything of value and probably the level did not extend far. About a mile up stream the vein shows calc spar where it crosses the beck. The eastern level has been driven through the thick boulder clay into the base of the mountain and the dump from it shows the vein to carry some good-looking galena in quartz.

At the 30 Fathoms Level a crosscut was put out eastwards to the main vein and drifts were started upon it in both directions. To the NNW very little ore was found and as the vein was very loose it let through a large amount of water from the beck. When it was realised that the best values were likely to be found in driving southwards a strong dam was built in this northern level to try to cut down the inflow. The Southern level was then continued, and by 1877 is said to have reached a point 150 fathoms from the shaft foot. It would therefore be some way in underneath the eastern shallow level. Old miners were emphatic that there was some good ore found in this long drive but no attempt was made to stope any of it. Some quantity of lead ore must have come from this work but there is only a record of 3 tons sold in 1870; it ran 66% lead and 10 ozs. silver per ton of ore. In any case this ore could not have come from the shaft and must be either out of nearby Dod End Mine or from the Shallow level.

Several parcels of ore were probably sent away in the period 1870-77 and Mr. Hedley Grenfell, son of Richard Grenfell, told me that he could remember a load of lead ore going away. He said it was loaded on to a four-wheeled flat waggon drawn by three horses whose harness was covered with polished brasses. He said he would be 10 years of age at the time so this was in 1876, but there is no written record of any ore raised.

The vein was porous and the amount of water draining down into the mine from the beck was a great problem as the pump had difficulty in keeping the flow in check. Hedley said that he spent many hours watching the great wheel turning steadily and the action of the sliding rods across the beck. It performed its task night and day without stopping. Mr. Grenfell, sen., proposed to lighten the load on the pump either by making a cutting in the rock behind the shaft (and so divert the beck away from the vein) or by carrying it along over the vein in large launders. He died in 1877 before either plan could be put into action. Very shortly afterwards the miners broke unexpectedly into a large cavity in the vein and had to flee for their lives. One of them told my father that the flood followed them up the ladders in the shaft and odd pieces of tackle floated about. Everything was lost. The pump would be kept going even though it was overwhelmed but it was unable to beat the water. and from that day forward no one has seen the 30 Fathoms Level. This disaster involved the complete loss of a brand new mine and so discouraged the shareholders of the Patterdale Mining Company that they decided to give up, so not only Myers Head but Eagle Crag also came to an end. Had they diverted the stream in one way or another the pump might have been able to clear the mine but this depended on its keeping on working, a situation that was only possible as long as the bucket lasted and the valves were kept clear.

At 400 yards NW. of the shaft a crosscut level was driven NE. into the low hillside at about 620 ft. above sea level. It appears to have been continued for some distance but no vein seems to have been encountered. It is possible that the level went over the main vein before the solid rock was reached. The 30 ft. Waterwheel was dismantled and re-erected at the Thornthwaite Mine, where it worked until this mine finally closed in 1920.

Myers Head vein had galena and also some zinc blende and a little chalcopyrite. Whether the blende is present in any quantity is not known; none was produced, but in those days the price was so low that unless it occurred abundantly the miners did not work it. No plan of the mine has survived. Mr. Hedley Grenfell died in 1934, aged 68 years, having spent almost all his working life at Greenside rising, from a washings boy to under manager. He was well respected by all who met him.

Helvellyn or Wythburn Mine.

THIS mine is the youngest of the group, not being worked before the 19th century. The existence of the veins was probably known long ago. The workings are in the upper part of the very steep Mines Gill, about one mile NE. of Wythburn church and are now filled in at the entrances. About 1840 the property was leased from Sir Ralph Vane, for one sixteenth royalty, by Mr. Henry Molyneux and Partners. They developed the veins by a system of adit levels. In 1857 they formed the Henry Helvellyn Lead Mining Co. Ltd., continuing the work until November, 1861, when a new company (Wythburn Lead Mining Co. Ltd.) was formed. Operations continued vigorously until 1872, when the undertaking was transferred to the West Cumberland Consolidated Mining Company. This firm continued working them until 1880, when Manchester Corporation took over Thirlmere for a water supply and brought mining to an end.

The country rock is the Volcanic Series of Borrowdale, containing here several felsite dykes, and both the veinstone and the country rock are extremely hard. Four veins have been discovered and tried—two NNE-SSW veins which hade east, the most westerly one being called the Blue Rock, and the other the Old Vein. These are joined by the SW-NE Brown Cove Vein, which dips SE. At the junctions there is said to have been some rich ore. The fourth vein runs E-W and dips steeply south, appearing to cut off the NNE-SSW veins, for they have not been found to the south. This E-W vein is thought to be the continuation of the Eagle Crag Vein at Patterdale but at Wythburn it is more or less barren.

The Arnison Level is the highest at about 2,100 ft and starts on the Old Vein. A short distance in from the level mouth the Brown Cove Vein is encountered and this continues on through to the Blue Rock Vein, which lies 15 yards to the west. At these junctions there was some rich ore. No. 1 Level is 150 feet vertically below Arnisons, and in this level the junctions produced good ore; they would be 400 feet in from the entrance.

100

No. 2 Level, which has been the main producing point, starts from the west side of the gill at exactly 1,750 ft., and has been driven at least 100 fathoms to reach the Brown Cove Vein junctions. The tip from this level indicates that it has been driven a considerable distance and that some of the felsite dykes have been met with. These are said to have deadened the veins in every case, which is rather different from the situation at Greenside Mine. There the dyke appeared to have little influence on the values of the vein, except where it was on both walls, when this vein was also barren. It is not known whether the dykes in Wythburn Mine were on one or both walls.

The veins are filled mainly with very hard veinstone, some quartz and barytes and the only mineral of importance is galena, which is mostly scattered throughout the vein. There is little solid ore. The Old Vein was the more productive and contained some highly argentiferous galena. Brown Cove Vein was not much worked away from the junctions but a good deal of driving was done on the Blue Rock which also yielded some ore.

Below No. 2 Level mouth a crude ore bin and grating house was set up, with a short tramway to the head of a 600 yards long double-acting 1 in 3 incline, by means of which the sorted ore was let down the mountainside to the dressing plant. This was laid out by the side of the gill at about 1,000 ft. and consisted of an ore bin with stone breaker followed by crushing rollers, trommel screens, jigs and buddles all propelled by a 30 hp. turbine. Water for this machine was stored in a masonry built dam higher up the gill. The whole was housed in a substantial building of which little remains. A short distance away, in a wood, is the well-built powder house, now used as a cabin by foresters.

No. 3 Level seems older than the others. It is 150 feet lower than No. 2, driven into the right bank of the gill as a crosscut to the Old Vein. Almost all the tip has been washed away by the beck and the entrance is closed so there is little evidence of its extent. No signs of ore can be found. Immediately above there has been a row of stone and slated buildings which housed the smithy and office. No. 4 Level is in the fir wood at 1,400 ft. and is a short crosscut to the E-W vein, presumably finding the junction with the Old Vein. No mineral is to be found on the tip.

It is said that, in all, 650 fathoms of levels were driven and only a few hundred tons of dressed galena were raised for a capital expenditure of £20,000. The veins, though hard, are said to be fairly wide and breaking the ground would present no problem today, nor would dressing the ore with modern machinery. It might now have been a profitable proposition. The mines can never again be worked however because of pollution to Thirlmere.

Tongue Gill Iron Mines.

ABOUT one mile NE. of Grasmere village are two ancient iron ore mines which have been opened on a haematite-bearing vein coursing NW-SE and dipping SW. through the Borrowdale Volcanic rocks. The easterly mine, called Fairfield, has been worked from two levels driven SE. along the vein but they have long ago been closed up at the mouths. From the lower level or adit there is a strong flow of water which is used to supply the village.

A large tongue of ground separates the becks at this point and the old packhorse road leads off to the left up Little Tongue Gill past the Providence Mine. Here three levels have been driven into the vein which is said to have been very wide and strong with up to 16 feet of solid haematite. In some of the stopes the ground was difficult to control and there was a fatal accident in this mine in the last working when a roof fall occurred.

Both mines were reopened and greatly worked in the early 17th century, when the ore was carted to Langdale for smelting. In 1874 Fairfield Mine produced 204 tons and Providence Mine 300 tons of haematite, which was carted the 14 miles to Windermere station. In all about 1,500 tons of ore was sent away during this last revival. Valued at only 14/– per ton it is easy to see why the venture failed.

Grasmere Mine.

THIS Elizabethan lead mine is high up Greenhead Gill and about one and threequarter miles NE. of Grasmere Village. A well-graded packhorse road leads steadily up the side of the steep gill to the site of the old dressing plant. The veins cross the beck in an E-W direction and are associated with a crush zone or shatter belt about 100 feet in width. Four small veins have been tried by opencasts and shallow shafts and a crosscut adit appears to have been some driven towards them from lower down the gill.

In 1569 the Elizabethan Dutchmen erected a stamping mill driven by a waterwheel with box buddles to dress the crushed ore. The stone buildings have been substantial but have gradually decayed so that now, 400 years afterwards, only the outlines of the walls remain.

The northern vein has been tried by a shaft which is still about 20 feet deep and is no doubt a great deal deeper. It has been cut out of the rock very neatly. Sinking it must have been a slow and difficult operation without the aid of explosives. Not much is to be seen of the two little opencasts, as they are now well grown over, and it is possible that short levels were driven along the veins from them. The southern vein, main source of ore, has been worked from a shaft called St. Benedicts which is across the beck opposite the old mill. A sample of low grade ore from here taken in 1939 ran:- Lead 2.65% Zinc 10.65% Silver 0.3 ozs. per ton of material.

The hopes of the Dutchmen were not fulfilled for they closed the mine down in 1573. It has never been reopened and so the remains are of interest as being truly Elizabethan.

Keswick Mining Field

THIS area contains many old mines which have been fully described by John Postlethwaite in his book *Mines and Mining in the Lake District*. The following information covers the period from the last edition of this work to present day and relates only to those mines which have been active during this time.

Threlkeld Mine.

AFTER the Borlase Company was liquidated in 1913 the mines were taken over by the late Mr. Anthony Wilson, who formed a fresh company called Threlkeld Lead Mines, Ltd. All the workings below the Horse Level were abandoned and flooded. Instead of trying to reopen them, Mr. Wilson decided to drive the level forward into new ground.

The level forehead had been standing for many years. When it was found that the main oreshoot was either cut off or died out at the "slide", no attempt was made to try to find further ore ahead of this interference until the late W. H. Borlase, jnr., became manager of the mines about 1909. Mr. Borlase started driving the level ahead following a barren but promising footwall and presently a heavy flow of water was cut. This nearly caused the loss of the mine as all the working faces were below the Horse Level and the water poured down into them, threatening to drown the pumps. Joiners and timber were hurriedly brought from Greenside Mine and boxes constructed to carry the water over the parts where the sole of the level had been cut away. The mine was saved although the 30 Fathoms Level remained flooded for some time afterwards. As the company's existence depended on raising ore from these lower levels it was decided to let the Horse Level forehead stand and this then was the position when Mr. Wilson took over.

After only driving by hand-boring for a few weeks the vein began to open into an oreshoot which eventually was 150 fathoms in length. It was stoped upwards some 23 fathoms to the Smithy Level (which had by then also been driven forwards into the ore) and to a considerable height above this level. These stopes were worked on the old system of cut and fill, with stone walled mill or hopper holes built up through the fill and extended as the roof of the stope advanced. Owing to the great length or height of these mills the force of the descending ore striking the sides gradually tore way parts of the walls and many of them were lost, disrupting the whole stoping system.

By 1928 the prices of lead and zinc had fallen below the mine's economic limit and the company decided to close down until an increase in prices would make it payable. The plant was maintained in good order ready to start up at short notice. In 1937 a London firm examined the workings and sampled the oreshoot along the floor of the Horse Level with results said to have been satisfactory, but nothing more was done. The plant was sold and dismantled during the 1939-45 war and since then the mine has been abandoned. Much of the timbering through long stretches of pack-filled ground along the Horse Level will have given way, and to reopen the mine would require much of this ground to be bypassed by driving a new parallel level in the solid rock.

The roof of the stopes are still a great distance from surface and the ore shows no sign of drying out. The face of the Horse Level is beyond the summit of Saddleback but still about 1,700 feet below the surface and there seems no reason why further ore should not be found if it was continued forward. In the old part of the mine, in the 30 Fathom Level stopes, there was much more zinc blende than lead ore and also an increase in iron pyrites which made separation by gravity means virtually impossible. The blende concentrates were very low grade and consequently of little value. The floatation process would now make it possible to produce first grade concentrates from this material and these long abandoned workings could be reopened and extended.

Brundholme Mine.

EARLY in this century Captain Borlase formed a syndicate and reopened the old workings which had been abandoned about 1880. The shaft was found to be 26 fathoms deep below the shallow adit level with levels off at 8, 16 and 24 Fathoms below the adit. Mr. Borlase pushed forward the 16 and 24 fathoms Levels both north and south along the vein and good pockets of lead ore were found; they were not sufficiently large to make the mine a paying proposition and before long the venture ended.

The late Mr. Bennett Johns of Keswick formed a company called Brundholme and Newlands Mines in 1918 and reopened these workings. The old shaft was very small and crooked and it was decided to sink a new one some distance to the south rather than to deepen the old one. The new shaft was connected to the 16 and 24 Fathom Levels and finally reached a depth of 35 fathoms below the adit.

Brundholme is a very wet mine, possibly because the veins run for some distance along the beck bottom, and there had been so much trouble with breakdowns of the engines and pumps that by then the company's resources were exhausted. Once again the mine was abandoned without any new lateral exploration having been done. There has been no further attempt to reopen the mine and all the machinery has long since been removed.

Goldscope Mine.

BRUNDHOLME and Newlands Mines Company, under Mr. Johns, reopened the Grand Level to the forehead some way beyond the famous lead workings and commenced driving the level forward along

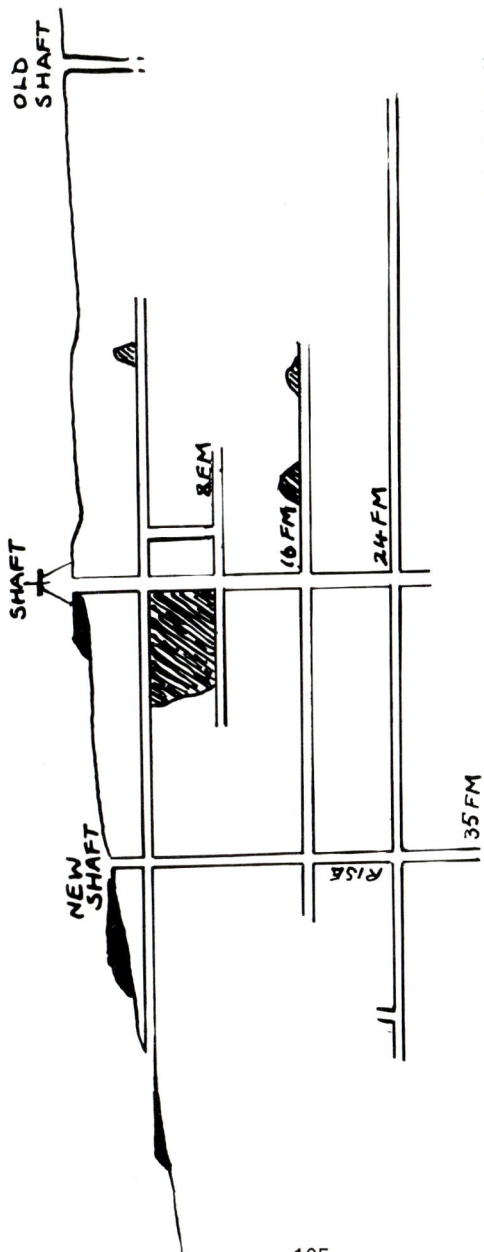

A longitudinal section of Brundholme Lead Mine.

SCALE :— 120 FT TO 1 INCH.

OLD SHAFT

SHAFT

NEW SHAFT

8 FM

16 FM

24 FM

35 FM

RISE

the copper vein in search of the junction with Sealbys lead vein. This lead vein was tried on a small scale long ago without much success but it was thought that the junction with the copper vein might reveal a rich deposit of lead ore similar to the Great Bunch. The level swung away to the north and presumably the miners were led away by following a false vein. It holed out to surface without finding any sign of Sealby's vein.

Although there are many known intersections of lead and copper veins in this area, only in the one which produced the Great Lead Bunch is there no dislocation of the copper vein. It seems reasonable to seek further bunches by following the Goldscope Lead vein towards other known copper veins even though the lead vein may be completely barren in the intervening stretches.

Dale Head: Long Work Mine.

AT the top of Newlands Valley an E-W copper vein known as the Long Work has been worked in a small way by several people, beginning with the Elizabethans. Mr. Johns commenced driving a crosscut in a S-W direction from the West side of Newlands beck opposite the small sheepfold about 1919. The vein was cut and water drained from an ancient open stope down to the level of the new crosscut, which was not however actually connected into it. Examination of the old working showed that it extended to an unknown depth below the crosscut, and as no machinery of any sort was available the venture was given up. Obviously the ancients had removed any worthwhile copper ore which had occurred above the horizon of the new level.

Recently this vein was examined by a Canadian geologist and an application was made for planning permission to try to establish a modern mine. The application was refused on amenity grounds.

Thornthwaite Mine.

THORNTHWAITE is an ancient mine which was restarted in 1873 by Capt. Wm. Francis. It was worked continuously until the slump in 1921, when it was closed down temporarily and finally abandoned when hopes of a rise in metal prices failed to materialise. During most of the years worked in this century its management was in the capable hands of the late Mr. Anthony Wilson.

The vertical Main Shaft is on the West side of the main Keswick-Cockermouth road about 200 yards south of the fourth milestone from Keswick and at 280 ft. There are levels off the shaft at 10, 17, 27, 37, 47, 65 and 85 fathoms. In the latter years of activity most of the work was done in the three lower levels. When the mine closed down there was still a considerable reserve of ground left to work above the 85 Fathoms Level and there seems no reason why the ore values should not continue to much greater depth.

Latterally in the workings to the south of the shaft, the ore ground ended at about 1,000 feet in a patch of disordered ground very similar to those seen in the veins at Threlkeld and Force Crag Mines. The 47

The Newlands Valley near Keswick.

Fathom Level was continued for some 500 feet further than the 85 Fathom Level but no workable ore was encountered. North of the shaft the farthest point reached is the Air Shaft which has been sunk a little way east of the main road but has now run in at the top. This shaft goes down to the 17 Fathom Level and is approximately 1,000 feet north of the Main Shaft. The vein is known to extend some way along the foot of Barf mountain, for an old trial there found a little lead, but the vein in the northern end of the mine levels is said to be restricted and obscure. In the orebearing parts the vein is large, sometimes reaching as much as 50 feet in width, but most of the values are concentrated towards the walls, with galena favouring the hangingwall side.

In the end, all the extensive dressing and other plant was removed and apart from the tailing heaps there is now very little to indicate that once in the not so very long ago here was a busy and worthwhile enterprise which employed an average of 100 men and boys, many of whom, living in Keswick, cycled to and from work daily. The closure of this mine after nearly 50 years of continuous operation was a local disaster.

Force Crag Mine.

THIS is the only mine in the area to have been worked after the depression, though barytes, not lead or zinc, was obtained. Barytes-working at Force Crag dated back to the 1860s when local men reopened the lead workings for this mineral, but by 1880 the venture had been given up as uneconomic for price reasons rather than lack of barytes. In 1905-6 Messrs. Lobb & Co., who were working Thornthwaite Mine, reopened the lead workings and erected a then modern dressing plant for the production of lead and zinc concentrates. It was impossible to separate blende and barytes, and as there was not sufficient galena present in the ore to allow the mine to be worked for lead alone, they closed down in 1909 after spending a great deal of money.

In 1912 the Coledale Mining Syndicate was formed and work restarted. This company installed one of the very first Elmore Floatation Plants, which separated clean blende from a mixed blende and barytes concentrate. Lack of water power and other inadequacies brought the mine once more to a stop in 1915. Braithwaite Mines Limited was formed in 1916 with a capital of £15,250, the principal subscribers being Mr. G. F. Wallace and Mr. C. E. Currie. Mr. Wallace acted as manager and proceeded to erect two suction gas engines to overcome the power shortage. A silex-lined ball mill was also installed to grind the coarse mixed concentrates from the jigs to feed the Elmore Plant, and several latest type concentrating tables were also erected. In 1919 an air compressor was set up to supply air to two rock drills. During the 1914-18 war the mine was worked with some vigour for the blende was in great demand by the Ministry of Munitions. The No. 0 Level was begun as a crosscut into the base of the mountain at 100 feet vertically below the No. 1 or Mill Level and by 1921 when the mine closed down it had been driven 350 feet in a NNW direction without finding any sign of the vein.

Looking down on Force Crag Mine.

A fresh company took up the lease of the mines in 1929 and drove the High Force crosscut in search of barytes. This level starts in the steep crags beside Pudding Beck at 1,710 ft. and after driving 520 feet a very fine vein of barytes was cut. Nothing further was done until 1939, when the Tampimex Oil Co. Ltd. took over and erected a ropeway down the mountain to carry the crude barytes to the millsite. Unfortunately in the long interval since 1921 all the former machinery and buildings had been removed and it was necessary to erect a new brick mill building which incorporated some of the original stone walls. A complete crushing and washing plant was installed by Messrs. Davison Ltd., of Hexham, the machines being belt-driven by an 80 h.p. Blackstone diesel engine. Several portable compressors were set up in a shed at High Force Level mouth and during the next seven years some 35,000 tons of high grade barytes was produced.

By 1947 almost all the worthwhile barytes-bearing ground had been removed to a depth of 80 feet below the High Force Level, which was the bottom level off an internal shaft. The 1947 blizzard brought the venture to an end, for by the time the miners were able to get into High Force Level all the shaft workings were full of water and all the pumps and other tackle had been lost. Messrs. LaPorte & Co. Ltd., bought up the lease and plant and drove a long incline from the old No. 3 Level to beneath the 80 ft. Level workings with a view to opening the barytes shoots at greater depth without having to pump or wind the ore up to High Force Level. After completing the incline and putting up a rise from the top of it to the 80 Ft. Level there was a recession in the barytes trade and they decided not to continue. All the plant and machinery was again sold up and removed.

Messrs. McKechnie Brothers Ltd., of Widnes, who were working the Caldbeck Barytes Mines, took up the property in 1960 and drove the No. 3 Level forward to beneath the ore shoots in High Force Level but without finding any workable barytes. Subsequently development and extraction of barytes from workings off the incline was begun and further connections made to High Force Level. The crude barytes ore was scraped down the upper part of the incline to a rise from No. 3 Level where it was drawn out to daylight by a battery-type locomotive in trains of six cars. Here the ore was dumped into a tipper lorry which carried it down the mountainside to the mill. A new plant had been set up in the building, only the shell of which remained, with each machine driven by individual electric motor, and the mill was capable of producing 100 tons of dressed barytes per week. Power for the motors was supplied by a 100 KW Dorman diesel set and there was a 50 KW standby alternator belt driven by a 120 hp. GMC diesel engine. Compressed air for the rock drills, fans and slusher hoist in the mine was from a 600 cu. ft. minute Joy Sullivan rotary compressor which had a Cummins diesel engine.

Development below the 80 Ft. Level was poor as the vein contained large amounts of wet clay with very little barytes; it was very difficult ground to work. In 1965 after clearing through a large fall the west face of High Force Level was seen for the first time and although it was completely barren there was nevertheless a good footwall and it was thought worthwhile to drive it on at least for a little way to see if there was any sign of further barytes. Presently the vein opened into a brand new shoot of barytes, which was slightly opened up for stoping, in which work up to 5 feet of excellent barytes was revealed. The vein in the face of High Force West level carries about 9 inches of barytes and there is no doubt that further shoots will be discovered if the level is ever advanced. On surface above this ground the rock near the hilltop is black shale in which the vein does not carry any mineral but immediately below this shale the rock is Skiddaw sandstone which is a very favourable host rock in which the barytes has been proved to be workable for at least 300 feet below the shale. Although very varied the general dip of the strata is gently down to the west and there remains some 1,500 feet to drive before the level would come out to surface.

In 1966 Messrs. McKechnie Brothers Ltd. withdrew from mining and gave up the manufacture of barium chemicals. All the plant and machinery was once again sold up by auction and the mine lease given up. A group of Canadian stockbrokers became interested in the possibility of reopening the old lead workings to see if it would be a payable proposition under present-day conditions. A new company called Force Crag Mines Ltd., was formed in Toronto with a share capital of 4,000,000 shares of $1 value.

A new entrance to the No. 1 Level has been driven through the loose shale and boulder clay and the level reopened and examined for 1,000 feet westwards. Eastwards the level has been reopened to the face and a little new ore found in a short drift eastwards. No. 0 Level has now been driven to below the eastern end of No. 1 Level and so far one new shoot of ore has been found. It is hoped that sufficient ore will be found below the old stopes in No. 1 Level to make it worthwhile to equip the mine with a modern floatation plant and diesel power units and bring it into production.

Coniston Mining Field

THE Coniston area has produced at least 75% of the total production of copper in the Lake District, the bulk of it from the Coniston group of veins. The Bonsor Vein produced about half the grand total. The mining area proper covers about 10 square miles of rugged mountainous country with peaks such as Coniston Old Man (2,633 ft.), Carrs (2,575 ft.), and Wetherlam (2,502 ft.). It has been thoroughly prospected and mined during the last 2,000 years.

The copper-bearing veins lie entirely in the volcanic rocks of the Borrowdale Series which here are of great thickness. Generally the veins have a north-westerly direction but their dip is very variable even in any one vein. The whole area is crossed by numerous north-south faults or crosscourses some of which are very powerful; they shift the veins for considerable distances. In the dozens of intersections of veins and crosscourses which have been met with in the mines almost all the throws have been to the left. Chiefly the veins are filled with stone rather than quartz and, apart from some dolomite, other gangue minerals are scarce. The copper occurs almost wholly as chalcopyrite with more rarely bornite. In some veins there are large quantities of iron pyrites, mispickel and some blende. In the deeper part of Bonsor Mine there was a great deal of magnetite.

The richest ore was usually found when the vein filling was a softer kind of black slate, and the veins were very poor when there was much quartz present. It had been realised from earliest times that there were two kinds of crosscourses—the large ones filled with douk or clay and a smaller, tighter type almost contemporary with the mineralisation, for it was found that very often when following a poor vein, as one of these crosscourses was being approached, there was a great improvement in ore values. The vein would widen out considerably as though it had been burst by the crosscourse.

Coniston's group of mines were worked extensively, especially in the 19th century, and they declined mostly because of the slump in copper prices towards the end of the century rather than because the veins were exhausted. There have been attempts to restart them, so far without success. The following is an account of the latter days of the mines in the area rather than details of past working in their heyday operations.

Coniston Water and the Old Man.

Coniston Mines.

The chief veins on this property are Bonsor, Triddle, North, Flemmings String, Paddy End Old and New, South, Belman Hole, Stephens, God's Blessing and Brimfell, most of which have been extensively worked.

Bonsor.

THIS vein was the largest producer and was stoped out almost completely for a length of a quarter mile; over a similar vertical extent it was reckoned that the vein throughout this colossal excavation would carry a rib of solid chalcopyrite eight inches wide.

On the surface the vein is seen to dip slightly south. About the drainage adit or Horse Level (which at the Old Engine Shaft is 300ft. below the surface) the vein is vertical, but as it goes deeper it gradually turns north. In the deepest workings, which are 205 fathoms below the adit, it is dipping flatly north though this factor does not appear to have had any effect on its width and mineral values. By 1895 the price of copper had fallen below the economic limits for these deep workings and for some time, as the mine went deeper, there had been an ever-increasing amount of magnetite in the ore, which made separation by gravity methods almost impossible. Reluctantly it was decided to stop pumping and allow the mine to fill.

As the water rose steadily any worthwhile pillars were removed, level after level, up to the Horse Level, which the water reached in 1900. The vein left in the bottom of the mine would pay because the separation of the minerals by the floatation process is now simple, but it would be prohibitive to pump out the water and rehabilitate the shafts and workings. The only work done in the mine this century, apart from the removal of some pillars at the New Shaft in 1908, was the attempt in 1954 to reopen the Horse Level through to Paddy End. After cleaning the level to a little west of Old Engine Shaft it was found that the whole great stope ar Cobblers Hole had collapsed. A new bypass level was driven in preference to trying to clear a way through the fall.

Some work had been done on Drygill vein in the olden days. This vein runs through the shaft at Horse Level; the old workings were cleared and the level driven forward about 50 fathoms, where it connected with the North Crosscut, which had been driven at the west side of the Cobbler Hole stope. It was hoped that much of the Horse Level would be passable towards New Engine Shaft but after a connection was made to the crosscut it was found that the extensive old stopes had collapsed. It was not possible to construct a safe way through other than by driving a new parallel level and the venture was given up on account of the cost. The price of copper had then fallen to under £250 per ton.

Mention should be made of the attempt to reopen the mines in 1912 by a French company associated with Minerals Separation Limited and formed the Coniston Electrolytic Copper Co. Ltd. After French mining engineers had sampled the various old dumps a plant was erected to recover the copper in a metallic state. The intention was to reopen the Paddy End Mine and to use the dump material to supply the plant until the mine began producing. The new machinery was set up at

Deep Level mouth at Bonsor Mine.

MINING IN THE LAKE COUNTIES

Bonsor, on the site of the old dressing plant and where there was a plentiful supply of waterpower available.

Briefly the layout was as follows: Dump material was brought by horse carts and fed by hand to a 20″ x 10″ stonebreaker driven by a 25 ft. diameter waterwheel. The crushed rock passed to a Wilfley Ballmill, which ground it to fine sand size, and this pulp was treated in an 8 cell minerals separation machine. Both the ballmill and the flotation machine were driven by a Gilkes pelton wheel. The copper-bearing froth was settled in large wooden vats and excess water allowed to drain away. It was then shovelled out and further dried on hot plates before being fed to a rake type roasting furnace. The roasted material was fed to a Patterson agitation tank, where sulphuric acid was added, and the resulting copper sulphate liquor was pumped to a batch of drum type electrolisers which produced thin copper sheets assaying 99.90% copper.

The actual copper recovered from the old dumps was much less than had been anticipated from the assays of the original sampling and the operation was closed down for the duration of the war in 1915 before any real work had been done at Paddy End. The plant was maintained in first-class condition throughout the war but afterwards work was not restarted and eventually it was dismantled and sent away to Greece.

Paddy End Mine.

IN 1954 work was done in an effort to explore the workings at the Horse Level and, if possible, to pump out the Hospital Shaft which goes down to the 20 Fathoms Level. The best chance of succeeding was to reopen Courteneys Crosscut at Paddy End and make a ladderway down the old South Shaft, which has been sunk down to the Horse Level on the South Vein. Tributers had stoped away the sides of the shaft and there was no opening into the Horse Level from the bottom of it.

In the latter part of the 19th century all available ore pillars were removed throughout the Paddy End workings and even in 1908 many of them were in a dangerous state, a condition which has deteriorated. Any future working of the veins could best be accomplished by making a completely new mine, sunk below the old workings for a start, with a good pillar of ground left in below them for safety. Whether there is sufficient copper available to cover such an outlay is unknown and there is no cheap way of finding out.

Triddle. No work has been done here since 1895 and the vein is believed to be worked out to 30 fathoms below the Horse Level. This was a busy mine in the 18th century when there was a stamp mill by the side of Red Dell beck.

Brimfell. The vein on Brimfell has been tried several times without much success. Mr. Day had a short level driven foward from an earlier working someway up from the base of the steep crag in 1908; very little ore was found. In 1930 Mr. Gnospelious had a level started

The floatation plant at Coniston Mine, 1914.

on the same vein about 20 fathoms above Day's Level; a considerable distance was driven but nothing of much value was found. Rock drills were used to drive both of these levels, air being brought by pipeline from the compressor at Bonsor Mine. In each case the miners were Shaws of Coniston.

God's Blessing.   This working, at the head of Red Dell, is thought to be even older than the Elizabethans. The vein is narrow but contained some rich ore which, as well as containing silver, is said to carry a little gold. In 1907 a few tons of ore were raised when an examination was made of the ancient workings. A flood some years ago almost obliterated them and the entrance of the short crosscut, which is called Coffin Level, is now closed.

Wetherlam Mine.

EARLY this century an effort was made by Mr. Day to develop the Wetherlam vein. A small dressing plant, consisting of a stone-breaker, three stamp Californian battery and a patent Bartlett concentrating table, was erected. The machines were driven by a paraffin engine and all were housed in a timber and sheeting building. Water for washing was pumped from nearby Crook beck by another engine. In the mine, which consists of a single level only, a small stope was started above the level roof and a winze was sunk in the floor for a depth of 13 fathoms. Compressed air was brought by pipeline over Hole Rake from Bonsor Mine and was used to work a pump and a small

Lark air engine for drawing the kibbles out of the winze. A horse gin was set up outside the level mouth and connected to a winch, at the top of the winze, by means of shafting using universal joints to navigate the corners and was presumably for use when no compressed air was available.

Operations were continuous for a few years only, and a small tonnage of concentrates running 9% copper was obtained. The vein in the foot of the winze is said to show six inches of solid chalcopyrite. An electrical prospecting system was tried out here for the Chile Copper Co. about 1906, and the numerous small diggings along the fellside were made in this connection. No large body of ore was discovered but several small veins were found and would appear worthy of further trial.

Tilberthwaite Mine.

THIS ancient mine lies at the head of Tilberthwaite Gill and has worked several copper bearing veins which run almost east and west across the beck, where they are cut off on the east side by a very powerful crosscourse. Early in the 1930s the Greenburn and Tilberthwaite Mining Co., Ltd., of London, cleared a way through the 500 fathoms long Deep or Horse Level and made a ladderway up through the old stopes to the surface. Exploration of the workings failed to find any ore of value, and nothing more was done.

Hellens Mine.

A strong vein running almost east and west across the southern side of Birk Fell is called Hellens Vein. It has been tried at various times without much success. In 1923 the Greenburn & Tilberthwaite Mining Co., Ltd., drove a shallow level eastwards along it into a low hill for a distance of about 30 fathoms and sank a sump six fathoms deep from the surface to connect with the level near the forehead. The vein was poor throughout this work and carried only small amounts of chalcopyrite, erubicite and some very dark-coloured zinc blende. All the work done was bored by hand by two Coniston miners, Joseph and John Hellen, after whom the vein is called.

A few years later a crosscut was driven north-westerly, some distance to the west of Hellens Level and rather lower down the hillside. It was planned to intersect Hellen's vein and also to find the junction with the nearby Borlase vein which runs NNE-SSW and is the only copper vein in the district with this direction. Rock drills were used in this crosscut which was in very hard rock but although some distance was driven nothing worthwhile was found.

A little way east of Hellens Mine, and on the same vein, is an ancient mine known as Walker's Works which is thought to have been started by the Elizabethans. There is a large old open stope and two little sumps which are mostly pre-gunpowder workings. A lower level some ten fathoms deeper is of later date and partly drains the stope but apparently it did not find any valuable ore. Much of the ore obtained from the old stope has been oxide.

Away to the west, where the vein climbs the hillside over into Greenburn side there is another Elizabethan level of short extent where a little ore has been obtained and on surface a little further west is an opencut made about 70 years ago where several tons of rich erubesite ore were obtained.

Man Arm Vein.   On the west side of Dry Cove Bottom (about 1,250 ft.) is an old copper trial which was reopened during the latter years of the Greenburn & Tilberthwaite Mining Co. Ltd., and a little ore raised. The vein proved to be very narrow and so far has proved uneconomic.

Pave York Mine.   This vein runs a little north of east along the northern slopes of Birk Fell and most of the mine workings are late 19th century. Three crosscut adit levels have been driven southwards to the vein at 20 fathoms apart, and a fair amount of stoping has been done. Most of the ore raised, especially near the surface, was copper oxide. This was crushed and treated with sulphuric acid in stone tanks, the copper then being recovered by precipitation. A little work was done in the lowest adit during this century.

Greenburn Mine. This was often called Great Coniston Mine and was the main mine of the Northern area. It was worked to a depth of 120 fathoms and was abandoned long ago when the fall in the price of copper made it uneconomic. At the end of its days an old miner remarked "That the shaft kibble took down £1 and brought up 10 shillings."

Terms Used in Mining

ADITS. Tunnels driven into a hillside to draw the water from a mineral deposit and to provide access and ventilation.

ADVENTURERS. Those who had shares in a mining venture.

ARCHING. Tunnel entrances and the parts driven through ground requiring support were often made permanent by building stone walls along the sides and arching them over similar to a bridge.

ASSAY. The percentage of metals in a sample of ore.

BACK. The roof of a working. The vertical extent between tunnels is said to be the backs.

BARYTES. Barium Sulphate.

BLASTING. Using explosives to break rock, usually by inserting a charge in a hole bored for the purpose.

BLENDE. Common name for Sphalerite. Also called Black Jack.

BOB. The horizontal beam of a Cornish type pump.

BRANCH. A vein which splits away from the main vein.

BUCKING IRON. A single-faced hammer used to coarsely pulverise ores prior to the invention of crushing rollers.

BUDDLE. An ancient device for concentrating ores crushed to sands size.

BUNCH. A small deposit of ore in a vein.

COUNTER VEIN. A vein running diagonally from one vein to another.

COUNTRY. The name given to rocks enclosing a mineral vein or deposit.

CROSS COURSE. A fault which crosses the vein being followed and usually displaces it.

CROSS CUT. A tunnel driven at right-angles to the direction of the vein.

CROSS VEIN. A vein running at or near right angles to the main vein.

CUT & FILL. A system of stoping in which the excavation made by removing the ore is filled up by waste material.

DEAD GROUND. A part of a vein containing no worthwhile mineral.

DEADS. Pieces of waste rock or useless mineral.

DRESSING. Cleaning ores ready for market.

DRIFT. A horizontal tunnel.

DRIVING. The operation of advancing a tunnel.

END. The farthest face of a working.

ENGINE SHAFT. The shaft or pit which has a winding engine and is usually the main one of the mine.

FATHOM. Six feet.

FAULT. A displacement of the veins and strata.

FOOTWALL. The wall or cheek of a vein on which you look down.

GALENA. Lead Sulphide: The common ore of lead.

GANGUE. Useless mineral in a vein.

GRATES. The small crude ore bins where the ore was sorted over a fixed grating.

GROUND. The country or stratum in which the vein is found.

HADE. The inclination of a vein or strata.

HANGING WALL. The wall or cheek of the vein which is uppermost.

HORSE. A large piece of barren rock between two branches of a vein.

HUSH. A cut made down a hillside by a stream of water assisted by men with picks until bedrock is reached and any veins laid bare.

JACK ROLL. A type of hand windlass used to draw kibbles from shafts and sumps.

JIG. A machine for separating mineral from waste rock by suspension in water.

JUMPER. A hand drill steel.

KIBBLE. A large bucket of wood or iron used to draw materials from shafts and sumps.

KNOCKINGS. Large pieces of mixed ore and waste rock.

LEADER. The joint indicating the true vein in barren ground.

LEATS. Water courses made along the sides of mountains to collect water for propelling waterwheels. They were given just sufficient fall to enable them to convey the required quantity of water without scouring the sides and bottom.

LEVELS. Tunnels to or along veins usually at set vertical intervals viz; 10, 20 and 30 fm. levels.

LODE. A regular principal vein.

NICK. Ancient name for a stope.

OLD MEN. All the miners of former times.

OPENCASTE. Where the vein has been excavated in daylight.

ORE SHOOT. The part of the vein containing workable ore.

PACK. Deads or waste used to fill up a worked out stope.

PARCEL. A heap of mineral dressed ready for market.

PIPE. A shoot of ore greatest in vertical extent.

PITMAN. One who looked after the pumps and shafts.

PLATING. Using long handled shovels to wash ore in a long box biddle.

POLL PICK. A single pointed pick much used in olden days.

PRICKER. A long tapered copper rod which had an iron eye handle and used in charging shot holes with gunpowder.

PROSPECTING. Searching for a mineral deposit.

PUSHER LEG. A telescopic device by which modern rock drills are supported and fed forwards as the hole is deepened.

QUARTZ. Silicon found in almost every vein and often the most common gangue mineral.

RISE. A shaft driven upwards.

TERMS USED IN MINING

ROCK DRILL. A machine for boring holes in rock usually driven by compressed air.

RUN IN. When the ground above gives way and fills the working.

SCRAPER. A thin iron rod one end of which is flattened and turned up similar to a little rake and used to clean out boreholes prior to charging them with explosives.

SETS. Timber frames generally of two legs and a cap or headtree used to support the roof and sides of a tunnel or level.

SHAFT. A pit either following a vein or sunk down beside it.

SHRINKAGE STOPE. A system of overhand stoping where the miners use the broken ore to stand on.

SHIFT. The period of work, usually 8 hours.

SIEVE. A screen made of wires or punched plate for sizing crushed ore.

SLIDE. A fault or slip which dislocates the vein.

SLIMES. The finest mud from the crushing machines.

SMIDDAM. The fine material from the grates.

SOLE. The floor of the working.

STAMPS. The original machines used for crushing ores.

STEMMING. The material and process of confining an explosive charge in a drill hole when preparing to blast.

STOPE. A working where the vein is cut away between levels. When working upwards it is overhand stoping. Downwards is underhand stoping.

STRINGER. A very narrow vein.

STULLS. Heavy horizontal timbers fixed from wall to wall in open stopes to hold back the hanging wall.

SUMP. A small shaft usually sunk from one level to another for access and to improve the ventilation.

TABLES. Machines for dressing ore crushed to below $\frac{1}{8}$" diameter.

TRIBUTE. A system of working in which the miners were paid for the dressed ore produced.

TUTWORK. These miners were paid for the actual ground excavated irrespective of whether it contained any ore or not.

TRUNKING. The process of separating clean mineral from crushed ore by washing in a long box buddle.

VANNER. A machine for treating slime ore invented late in the 19th century.

VEIN. A fissure in the rock containing minerals.

VUGH. A natural cavity in a vein often lined with beautiful mineral specimens.

WHIM. A machine used for drawing large kibbles out of a shaft and usually worked by a horse.

WINZE. A small shaft but larger or longer than a sump.

A Selective
Index